JN092584

航空無線通信士

英語簡易辞書

発　行

一般財団法人情報通信振興会

ま え が き

航空機関士として長年勤務し、また、航空無線通信にかかわる講習を長年担当してきたその経験・知識を元に、航空無線通信の人材育成に用いる資料の一つとしてこの〔航空無線通信士英語簡易辞書〕を作成いたしました。

簡易辞書は、英和編、和英編及び略語集からなります。これらに収録された用語は、航空無線通信に必要とされるものに限られています。また各語の説明も航空無線通信用語としての説明としています。

英和編は、通常の英和辞書と同様の要領で利用してください。和英編は、日本語の用語と対応する英語のみですから説明を要する場合には英和編に転じて内容を把握してください。また、略語集には、航空交通管制の業務や機関および通信用語などの略語を収録いたしました。

不十分な点の多い辞書ではありますが、多くの方々のご利用により航空無線通信の基盤の充実に資することができれば幸いです。なおこの簡易辞書が〔航空無線通信士用の英語標準教科書〕の付属資料として利用され、航空無線通信の人材育成に貢献できることを希望しています。

平成22年１月
令和３年３月（改訂）
編者

目 次

第1章　英和辞書……………………………1

第2章　和英辞書……………………… 127

第3章　略語辞書………………………201

第1章

英和辞書

凡　例

〔自〕：動詞（自動詞）
〔他〕：動詞（他動詞）
〔可〕：名詞（可算）
〔不〕：名詞（不可算）
〔複〕：名詞（複数形）
〔単〕：名詞（単数形）
〔代〕：代名詞
〔助〕：助動詞
〔形〕：形容詞
〔副〕：副詞
〔前〕：前置詞
〔接〕：接続詞

A

A

abate	
〔自〕和らぐ 軽減する	嵐は和らいだ。 The storm has abated.
〔他〕和らげる	離着陸の飛行方式は、空港周辺の騒音を軽減する。 Takeoff and landing procedures abate noise around the airport.
abatement 〔不〕減少	騒音軽減方式 noise abatement procedure
abbreviate 〔他〕 (語･句を) 略して書く	日本標準時は、JSTと略される。 Japan standard time is abbreviated to JST.
abbreviated term 略語	航空交通管制は、略語ではATCである。 Air traffic control is ATC in abbreviated term.
abeam 〔副〕～の真横に	当機の位置は、局の真横である。 Our position is abeam the station.
abort	
〔自〕中断する	パイロットは、離陸を中断した。 The pilot aborted takeoff.
〔他〕中断させる	離陸は中断された。 The takeoff was aborted.
above 〔形〕	
上述の 上の	上述の手順に従いなさい。 Comply with the above mentioned procedure.
～より高く	航空機は、雲の上を飛行した。 The aircraft flew above the clouds.
absolute 〔形〕 絶対的な	絶対高度は、電波高度計で計られる。 The absolute altitude is measured by a radio altimeter.
absolutely 〔副〕絶対的に	操縦室のドアは、絶対に開けてはならない。 The cockpit door must be absolutely

kept closed.

A

accept

〔自〕受け入れる

無線局は、遭難呼出しを絶対優先的に受け付ける。

Radio stations accept distress calls with absolute priority.

〔他〕受諾する

管制官は、パイロットの高度変更の要求を受諾した。

The controller accepted pilot's request of flight level change.

access

〔不〕出入り 接近 アクセス

操縦室への立入りは禁止されている。

Access to the cockpit is prohibited.

〔他〕接近する アクセスする

操縦室扉には、誰も接近してはならない。

No one can access the cockpit door.

accordance

〔不〕一致 調和

in accordance with ～

～に従って

パイロットは、レーダーベクターの指示に従って降下した。

The pilot descended in accordance with the instructions of radar vectoring.

accuracy 〔不〕正確さ

VOR/DMEによる航法の正確さは、高く評価されている。

Navigation accuracy by use of VOR/DME is highly appreciated.

accurate〔形〕正確な

正確な高度　accurate altitude

accurately〔副〕正確に

速度を正確に維持する

accurately maintain the speed

acknowledge〔他〕認める

パイロットは、急な風向の変化を認めた。

The pilot acknowledged sudden wind direction change.

acknowledgement

〔不〕承認

受信証　acknowledgement of receipt

acquire 〔他〕得る 捕捉する

管制官は、目標機影をレーダーで捕捉した。

The controller acquired the target aircraft on his radar screen.

act

〔可〕行為
不注意な行為が、ミスの原因となる。
An inadvertent act causes faulty operation.

〔自〕行動する
直ちに行動する immediately act

action
〔不·可〕活動 行動
迅速な行動 prompt action

active 〔形〕活動中の
使用中の離陸滑走路
active takeoff runway

activate 〔他〕作動させる
防氷系統を作動させる
activate the anti-ice system

activity 〔不〕活動
火山は活動している。
The volcano is in activity.

add〔他〕加える
長時間飛行のため、パイロットを一人加える。
Add one pilot for a long flight.

addition〔不·可〕追加
荷物の追加は制限される。
Addition of cargo is restricted.

additional〔形〕追加の
追加の燃料が必要である。
Additional fuel is required.

additionally
〔副〕追加的に
荷物が追加された。
Cargo was additionally loaded.

additive
〔可〕〔形〕添加物
添加物は、注意深く取り扱われるべきである。
Additives should be carefully handled.

address

〔可〕宛名 住所
自宅の住所 home address

〔他〕～にあてる
離着陸中のパイロットに対しては、通報は送らない。
No message is addressed to the pilots taking off or landing.

adjust

〔自〕順応する
パイロットは、新機種にすぐに順応した。

A

	The pilot has soon adjusted himself to the new type aircraft.
〔他〕調節する	周波数を調節する adjust the frequency
adjustment 〔不〕調整	電圧を調整する make an adjustment of voltage
advance	
〔可〕〔不〕進歩	科学は、進歩した。 Science has made advances. 技術の進歩　advance in technology
〔自〕進む	夜が更ける。　The night advances.
〔他〕早める	出発時刻を早める advance departure time
〔形〕事前の	事前の通告　advance notice
adverse 〔形〕不利な 反対の	不利な気象条件（悪天候） adverse weather conditions
adversely 〔副〕不利に 逆に	風が逆に吹いていた。 Wind was adversely blowing.
advice〔不〕助言 勧告	適切な助言　appropriate advice
advise 〔自・他〕忠告する	機長は、パイロットに減速するよう忠告した。 The captain advised the pilot to reduce speed.
aerodrome〔可〕飛行場	飛行場には種々の無線施設がある。 Aerodromes are equipped with various radio facilities.
aerodrome control 飛行場管制	飛行場管制は、タワーが行う。 Aerodrome control is conducted by the tower.
aerodrome reference point　飛行場標点	飛行場標点から半径９kmの範囲は、管制圏である。 The area of 9km radius from the aerodrome reference point is the control zone.
aeronautical〔形〕航空の	航空技術の発達は目覚ましい。 Technical advance in aeronautical engineering is remarkable.

aeronautical information publication（AIP）	航空路誌（AIP）
aeronautical administration communication (AAC)	運航業務通信 （AAC）
aeronautical operations communication (AOC)	運行管理通信 （AOC）
aeronautical public communication (APC)	航空公衆通信 （APC）
aeronautical mobile service	航空移動業務
aeronautical station	航空局 （無線局）

---

affect〔他〕影響を及ぼす

気象条件は、運航に影響を及ぼす。
Weather conditions affect flight operation.

---

affirmative
〔不〕肯定
〔形〕肯定的な

肯定する reply in the affirmative
パイロットは、肯定的な応答を送信した。
The pilot transmitted an affirmative response.

---

affix 〔他〕添付する

運航管理者は、飛行計画に署名する。
A dispatcher affixes his signature to the flight plan.

---

again 〔副〕再び

パイロットは、離陸を再び断念した。
The pilot aborted takeoff again.

---

agency 〔可〕政府の機関

タワーは、一つの政府機関である。
The tower is one of the government agency.

---

ahead 〔副〕前方へ

先へ進む　go ahead

---

aid
〔不〕援助
〔他〕支援する

航法援助施設　navigation aid facility
ATCは、運航を支援する。
ATC aids flight operations.

---

air
〔不〕空 大気

グライダーは、空を滑空している。
The glider is soaring in the air.

A

〔形〕空の 航空の	空の汚染　air pollution
air route surveillance radar	航空路監視レーダー　ARSR
air space	空域
air to air	空対空
air traffic	航空交通
air traffic control	航空交通管制　ATC
air traffic service	航空交通業務　ATS

airborne 〔形〕

機上の 離陸した（状態）	機上の無線装備 airborne radio equipment 機が浮揚した。 The aircraft has become airborne.
airborne collision avoidance system (ACAS)	機上衝突防止装置（ACAS）

aircraft 〔可〕（単複同形）	航空機
aircraft station	航空機局
aircraft earth station	航空機地球局
aircraft operating agency	航空機運航機関

airfield 〔可〕　飛行場

airline 〔可〕　航空会社

airliner 〔可〕　定期旅客機

airport 〔可〕	空港
airport surface detection equipment	飛行場面探知レーダー　ASDE
airport surveillance radar	飛行場監視レーダー　ASR

airway 〔可〕　航空路

alert

〔可〕警報 〔不〕警戒 〔形〕警戒している(状態) 〔他〕警告する	パイロットは、不測の事態に対して警戒している。 Pilots are alert to the unexpected danger. 管制官は、パイロットに悪天候の警告を出した。

	The controller alerted pilots of significant weather.
all station call 一括呼出し	管制官は、一括呼出しで気象情報を配布した。 The controller delivered weather information through all station call.
allow 〔他〕許す	最大離陸重量のため、追加燃料の搭載は出来ない。 Additional fuel loading is not allowed because of the maximum takeoff weight.
along 〔前〕～に沿って	パイロットは、新しいルートに沿って飛行した。 The pilot flew along the new route.
alphabet 〔可〕 アルファベット	英語のアルファベットには、26文字ある。 There are 26 letters in the English alphabet.
alphabetical 〔形〕 アルファベットの	アルファベットの文字 an alphabetical letter
although 〔接〕～だけれども	風は強かったけれども、飛行は快適であった。 Although the wind was strong, the flight was comfortable.
altimeter 〔可〕高度計	飛行機の高度計は、気圧高度計である。 The altimeter of an aircraft is the barometric altimeter.
altimeter setting 高度計の補正	14,000フィート以下の高度計補正には、QNHを使う。 QNH is used for altimeter setting below 14,000 feet.
altitude 〔可・不〕高度	ジェット機は、高高度を飛行する。 Jet aircraft fly high altitude.

A

amateur 〔可〕アマチュア　アマチュア音楽家
an amateur musician

amplify 〔他〕増幅する　入力信号は、増幅される。
Incoming signal is amplified.
amplification 〔不〕増幅　増幅方式　amplification method
amplifier 〔可〕 増幅器　受信機には、多くの増幅器がある。
A receiver has many amplifiers.

amplitude
〔不〕大きさ 振幅
amplitude modulation 振幅変調　振幅変調は、一般的な変調方式である。
Amplitude modulation (AM) is a common modulation method.

announce 〔他〕発表する　機長は、離陸中断を知らせた。
The captain announced aborted takeoff.
announcement 〔可·不〕 発表　告知

annul 〔他〕無効にする　以前のクリアランスは、取り消された。
The previous clearance has been annulled.

another 〔形〕別の
〔代〕もう一つの物
もう一人の人
別の制限事項も考慮しなければならない。
Another limitation must be considered.

answer
〔可〕回答　回答は、まだ来ない。
The answer has not arrived.
〔自〕答える　トランスポンダーは、自動的に返答する。
The transponder automatically answers.
〔他〕答える　受信局は、直ちに返答した。
The receiving station immediately answered.

antenna 〔可〕アンテナ　送信アンテナは、電流を電波に変換する。

A transmitting antenna converts electric current into a radio wave.

any 〔形〕どんな～でも

着陸は、問題なく行われた。

Landing was made without any difficulty.

anyone 〔代〕誰でも

操縦は、誰でも習える。

Piloting is easy that anyone can learn.

apart 〔副〕離れて

apart from～ ～から離れて

制御ユニットは、システムとは別に操縦室内にある。

The control unit is in the cockpit apart from the system.

apply

〔自〕適合する

この練習は、初心者には適合しない。

This practice does not apply to the beginners.

〔他〕適用する

通信には、標準用語を適用する。

Standard phraseology is applied to communication.

applicable 〔形〕適用できる

この練習は、初心者に適用できる。

This practice is applicable to the beginners.

application 〔不〕適用

最新技術の適用

application of advanced technology

approach

〔不〕（航空機の着陸地点への）進入

タワーは、滑走路33Rへの進入許可を発行した。

The tower issued approach clearance to the runway 33R.

approach control 進入管制

進入管制は、レーダー誘導で行われる。

Approach control is conducted by radar vectoring.

approach course 進入経路

標準進入経路は、STARに記載されている。

A

The standard approach course is mentioned in STAR.

---

appropriate〔形〕適切な
機長は、操縦士に適切な助言を与えた。
The captain gave the pilot an appropriate advice.

appropriately〔副〕適切に
管制官は、到着機の間隔を適切に調節する。
The controller appropriately adjusts separation between arrival aircraft.

---

approve〔他〕承認する
管制官は、パイロットの位置通報の修正を承認した。
The controller approved pilot's correction of position report.

---

approximate

〔形〕おおよその
VOR局上空のおおよその時間
approximate time over the VOR station

〔自〕接近している
機の位置は、NDB局に近いはずである。
The aircraft position must be approximate to the NDB station.

〔他〕接近する
速度が限度に接近している。
The speed approximates the limit.

approximately〔副〕おおよそ
約10分出発が遅れた。
Departure delayed approximately 10 minutes.

---

area〔可〕区域 場所
飛行場周辺の場所は、管制圏である。
The area around the airport is the control zone.

area control　空域管制
空域管制は、ACCが管轄する。
Area control is directed by the area control center (ACC).

area control center
航空交通管制センター (ACC)

area navigation 広域航法
広域航法により運航効率が向上した。
Area navigation (RNAV) has

	improved flight efficiency.
arrive 〔自〕到着する	航空機は、10分遅れで到着する。 The aircraft arrives at 10 minutes delay.
arrival 〔不〕到着	到着予定時間 estimated time of arrival (ETA)
as soon as possible 出来るだけ速やかに	機長は、飛行終了後出来るだけ速やかに到着報告をしなければならない。 The captain must make arrival report as soon as possible after completion of the flight.
assign 〔他〕割り当てる	無線局には多くの周波数が割り当てられている。 Many frequencies are assigned to a radio station.
assignment 〔可〕割当て	機長は、各乗員に職務の割当てを行った。 The captain made duty assignment to each crew member.
assist 〔自〕手伝う 援助する 〔他〕手伝う 援助する	機長は、離陸時パイロットの操縦を手伝った。 The captain assisted the pilot in his maneuver at takeoff.
assistance 〔不〕援助	効果的な援助 effective assistance
associate 〔自・他〕関係づける	急ぐことには危険が伴う。 There is a risk associated with hurry.
attempt 〔可〕試み 企て	通信設定の試みは、失敗であった。 Attempt of contact was unsuccessful.
〔他〕試みる 企てる	パイロットは、突風の中着陸を試みた。 The pilot attempted to land in a gusty wind.

attenuate

〔自〕衰える　突風は弱まった。
The gusty wind has attenuated.

〔他〕弱める　小型機は、先行機の後流乱気流が弱まってから離陸した。
The small aircraft took off after the wake turbulence of the preceding aircraft has been attenuated.

attenuation 〔不〕減衰　高周波の電波の減衰が目立つ。
Attenuation of high frequency radio waves is prominent.

authorize 〔他〕認可する　認可された乗員が通信を実施する。
An authorized flight crew will conduct communication.

authority 〔可·不〕権限　指定された者のみ立ち入りの権限を有する。
Only the assigned person has authority of entering.

automatic

〔可〕自動操作装置　FBWによる操縦装置は、自動操作装置である。
Flight control of FBW is an automatic.

〔形〕自動的な　自動着陸装置は、機の進入経路を精密に制御する。
The automatic landing system precisely controls the path of an approaching aircraft.

automatically 〔副〕自動的に　航法情報は、IRSにより自動的に把握される。
Navigation information is automatically obtained by IRS.

automatic dependent surveillance (ADS) 自動従属監視　自動従属監視システムによって洋上における運航の自動的監視が、可能になった。
ADS system has enabled automatic monitoring of oceanic flight operations.

automatic direction finder (ADF) 自動方向探知機	ADFは、NDB局への方位を求める機上の受信機である。 ADF is an airborne receiver to find the direction to a NDB station.
available 〔形〕利用可能な	運航に必要な情報は、飛行中常時ATC通信を通じて入手可能である。 Information necessary for flight operation is always available during flight through ATC communication.
avoid 〔他〕回避する	機は、右旋回で雲を避けた。 The aircraft avoided the cloud by a turn to the right.
avoidable 〔形〕回避可能な	前方の雲は、回避可能である。 The cloud ahead is avoidable.
aware 〔形〕 ～に気づいて ～を知って	航空無線従事者は、ICAOの標準用語を熟知しているべきである。 An aeronautical radio operator should be well aware of the ICAO standard phraseology.
azimuth 〔可〕 （天文の） 方位･方位角	レーダースクリーン上の方位は、アンテナの角度に同調している。 Azimuth on the radar screen is synchronized with the antenna's direction.

B

band 〔可〕バンド 周波数帯	通信電波の周波数帯は、八つに分かれている。 The frequency band of radio waves used for communication is divided into eight categories.
base 〔可〕基部 底	雲高（シーリング）は、地表面と雲底の間の縦の距離である。 Ceiling is a vertical distance between the surface and the cloud base.

B

〔他〕～に基礎を置く	運航の基本は、安全にある。 Flight operations are based on flight safety.
beacon 〔可〕標識 ビーコン	無線標識は、航路上に設置されている。 Radio beacons are located along the route.
bear 〔他〕保つ **bear in mind** 心に留めておく	私は、経験したことを心に留めておくべきであった。 I should have borne in mind what I had experienced.
bearing 〔可〕 （磁方位の）方位 方位角	VOR局に対する方位は、コースを意味する。 Bearing to a VOR station means that it is a course to it.
begin	
〔自〕開始する 始まる	飛行前ブリーフィングは、もうすぐ始まる。 Preflight briefing will begin soon.
〔他〕開始する 始める	飛行前ブリーフィングを始めよう。 Let's begin the preflight briefing.
beginning 〔可〕初め	初めから終わりまで from beginning to end
below 〔前〕～より以下に 〔副〕以下の	250ノット以下に減速せよ。 Reduce speed to below 250kt. 下の表を参照　refer the table below
beside 〔前〕～のそばに	スイッチは、計器のそばにある。 The switch is beside the indicator.
besides 〔副〕そのうえ	雨の上にさらに突風まで吹いてきた。 Gusty wind started to blow besides rain.
between 〔前〕～の間の	航空機と地表面の間の縦の距離は、高度である。 The vertical distance between an

aircraft and the surface is the altitude.

beyond
〔前〕～の範囲を超えて
向かい風は、予報以上である。
Headwind component is beyond the forecast.

blind
〔形〕盲目的な　盲目的な送信　blind transmission
〔副〕盲目的に　盲目的に送信する　transmit in blind

board 〔他〕乗込む
on board　機内に
乗客は、すべて搭乗した。
All passengers are on board.

bore-scope〔可〕
ボアスコープ
ボアスコープで内部を検査しなさい。
Examine by a bore-scope.

both 〔形〕両方の
送信機、受信機ともに不具合であった。
Both transmitter and receiver were malfunctioning.

brief 〔形〕簡潔な 短時間の　簡潔な通信　brief communication
briefly
〔副〕簡単に 手短に
彼は、手短に状況を説明する。
He briefly explains the situation.

broadcast
〔可〕放送 放送番組
broadcasting〔不〕放送
broadcasting station
放送局
野球の放送
a broadcast of a baseball game
ラジオ放送　radio broadcasting

busy 〔形〕多忙な 忙しい
パイロットは、アプローチ中は忙しい。
Pilots are busy during approach.

C

calculate
〔自･他〕算出する
計算する
FMSは、航法データを自動的に算出する。
FMS automatically calculates navigation data.

calculation〔可･不〕計算
コンピューターによる計算は、速く、

C

	かつ、正確である。 A calculation by a computer is rapid and accurate.
calibrate 〔他〕（計器などの）目盛りを調整する	パイロットは、高度計の目盛りを調整する。 A pilot calibrates the altimeter by the new altimeter setting.
call 〔他〕呼び出す	航空機局が最初に航空局を呼び出す。 An aircraft station initially calls the aeronautical station.
call sign 呼出符号 コールサイン	各航空機には、呼出符号がある。 Each aircraft has a call sign.
call system 呼出装置	SELCALは、無線通信の自動呼出装置である。 SELCAL is an automatic call system of radio communication.
calling procedure 呼出手順	標準呼出手順が、ICAOによって提供されている。 Standard calling procedures are provided by ICAO.
calling station 呼出しをしている局	呼出しをしている局は、他の局の通信を妨害してはならない。 A calling station must not cause interference to the other stations' communication.
station called 呼び出された局	呼び出された局は、受信証を以って応答する。 The station called responds with an acknowledgement of receipt.
cancel〔他〕取消す	パイロットがゴーアラウンドした場合には、着陸許可は自動的に取消される。 Landing clearance is automatically cancelled when the pilot executed go-around.
cancellation 〔不〕取消し	クリアランスの取消し cancellation of a clearance

capable 〔形〕～が出来る	新しいコンピューターは、より早く問題を解決することが出来る。 The new computer is capable of much faster problem solution.
care 〔可·不〕注意	取扱い注意　Handle with care!
careful 〔形〕注意深い	無線機は、注意深い取扱いが必要である。 Careful handling is required for a radio equipment.
carefully 〔副〕注意深く	パイロットは、離陸推力を注意深く決める。 Pilots carefully determine the takeoff thrust.
careless 〔形〕不注意な	不注意な操作は、予期しない問題に至る。 Careless operation results in an unexpected problem.
carelessly 〔副〕不注意に	パイロットは、不注意にコースから外れた。 The pilot carelessly deviated from the course.
carry〔他〕運ぶ 運搬する	ジェット旅客機は、300人以上の乗客を運ぶ。 A jet airliner carries more than 300 passengers.
carrier wave 搬送波	搬送波は、信号波によって変調される。 A carrier wave is modulated by a signal wave.
categorize〔他〕分類する	航空機は、幾種類かに分類されている。 Aircraft are categorized into some types.
category 〔可〕区分 種類	UHFは、マイクロウェーブの分類に入る。 UHF is in the category of micro-waves.

C

C

cause

〔不〕原因 理由 — 到着遅延の原因は、向かい風であった。
The cause for arrival delay was high headwind component.

〔他〕原因となる — 夜間効果で方位情報が信頼できなくなった。
Night effect caused unreliable bearing information.

centerline 〔可〕中心線 — 通常の側面上の進入経路は、滑走路中心線の延長である。
Normal lateral approach course is the extension of the runway centerline.

certain 〔形〕確かな — 磁気嵐が通信不能をもたらしたことは確かだ。
It is certain that the magnetic storm caused communication failure.

certainty 〔不〕確実 確信 — 正確な航法の確信がある。
There is certainty of accurate navigation.

certify 〔他〕証明する — 国の機関が、無線従事者の資格を証明する。
Government agency certifies radio operator's qualification.

certificate 〔可〕証明書 免許状 — 無線従事者の免許は、適任証書の一つである。
The radio operator's license is a certificate of competency.

change

〔可〕変更 — 運用限界の変更が発行された。
Change of operational limit has been issued.

〔不〕変化 — 風向の変化が速かった。
Change in wind direction was rapid.

〔自〕変わる 変化する — 天候は、急激に変化した。

〔他〕変える 変化させる	The weather rapidly changed. アンテナは、電流を電波に変える。 An antenna changes electric current into a radio wave.
channel 〔可〕チャンネル	チャンネルは、割当てられた周波数帯である。 A channel is an assigned frequency band.
chapter 〔可〕章	この本は、六つの章から成る。 This book consists of six chapters.
character 〔可〕文字 符号 **characteristic** 〔形〕特徴的な 特有の **characteristic frequency** 特性周波数	登録番号の4文字が、機のコールサインに使われている。 Four characters of a registration number are used in the aircraft's call sign. 例えば、搬送波周波数は、特性周波数として挙げることができる。 A carrier frequency may, for example, be designated as the characteristic frequency.
check 〔他〕調べる	運航管理者は、広範囲の気象を調べる。 A dispatcher checks weather conditions of wide area.
circumstances 〔複〕状況 環境	対流圏内の環境は、電波の伝搬には良くない。 The circumstances in the troposphere are unfavorable to radio waves propagation.
Civil Aviation Law (CAL) **Civil Aviation Regulation (CAR)**	航空法 航空法施行規則
clarity 〔不〕明快 透明さ	パイロットの気象報告では、はっきりした説明が重要である。 Clarity of expression in pilot's

C

weather report is important.

---

classification
〔可・不〕分類 類別 — 航空機は、いくつかの分類によって区別される。
Aircraft are discriminated by various classifications.

classify 〔他〕分類する — 通常、雲は、10種類に分類される。
Usually, clouds are classified into 10 types.

---

clear 〔他〕許可を与える — 機には、離陸許可が与えられている。
The aircraft is cleared for takeoff.

〔形〕澄み切った — 澄み切った青空　clear blue sky

clear air turbulence
晴天乱気流 — 晴天乱気流は、ジェット気流の近くにある。
Clear air turbulence (CAT) exists beside jet stream.
夕空に飛行機がくっきりと見えた。
An aircraft was seen clear against the evening sky.

clearly
〔副〕明瞭に はっきりと — 各語は、はっきりと発音されるべきである。
Each word should be clearly pronounced.

---

clearance 〔不〕クリアランス（航空機に対する）許可 — タワーは、突風が強いので離着陸許可を出すことは出来ない。
Tower can't issue takeoff or landing clearance because of strong gusty wind.

clearance delivery section
管制承認伝達席 — 管制承認伝達席は、タワーに位置する。
The clearance delivery section is located in the tower.

---

climb 〔自·他〕上昇する
〔可〕（単数形） 上昇 — この飛行機の上昇性能は、素晴らしい。
The climb performance of this aircraft is excellent.

---

closure 〔可·不〕閉鎖 — 滑走路34Rは、クローズされる。

Closure of the runway 34R will take place.

closure rate 接近度

交通の邪魔になる航空機の接近度は、危険というほどではない。

Closure rate of the hazardous aircraft is not critical.

---

cloud 〔可・不〕雲

パイロットは、発達中の雲に関心を払う。

Pilots are interested in a developing cloud.

cloud base 雲底

地表面と雲底の間の距離は、雲高である。

Ceiling is the distance between the surface and the cloud base.

---

coarse 〔形〕きめの粗い

きめの粗い航法は、許されない。

No coarse navigation is allowed.

---

coast 〔可〕沿岸

日本の沿岸　the coast of Japan

coast line 海岸線

coastal 〔形〕沿岸の

沿岸地域には、VHF通信が適用される。

VHF communication is applied to the coastal area.

---

coaxial cable
〔可·不〕同軸ケーブル

同軸ケーブルは、UHF以下のフィーダーとして広く使われる。

Coaxial cable is widely used as a feeder of UHF and lower frequencies.

---

cockpit 〔可〕操縦室

操縦室扉は、飛行中閉めて施錠してある。

The cockpit door is closed and locked during flight.

---

code
〔可〕符号 記号

運航は、それぞれの飛行便名を含む符号で識別される。

Each flight is identified by a code that includes the flight number.

C

〔他〕コード化する	エンコーダーは、機のSELCAL記号をコード化する。 The encoder codes the aircraft's SELCAL code.
collision 〔不〕衝突	航空交通業務は、航空機の衝突を防止する。 Air traffic service (ATS) prevents collision of aircraft.
column 〔可〕円柱状のもの **control column** 操縦桿	パイロットは、離陸のために操縦桿を引いた。 The pilot pulled the control column for takeoff.
combination 〔可〕組合せ	電波は、電界と磁界の組合せである。 Radio wave is a combination of an electric field and a magnetic field.
command 〔可〕命令 〔他〕命令する	 命令に従う　comply with an order MAYDAYは、通信の沈黙を命令する。 MAYDAY commands communication silence.
commence 〔他〕開始する	離陸を開始する　commence takeoff
common 〔形〕共通の 共有の	通信には、共通の理解が必要である。 In communication, a common understanding is required.
communication 〔不〕通信	ATC通信は、通常英語で行われる。 Normally, English is applied to ATC communication.
communication establishment 通信設定	送信すべき通報を持っている局が、通信設定を開始する。 The station having a message to transmit will initiate communication establishment.
communication failure 通信不能	トランスポンダーコード7600は、通信不能を意味する。 The transponder code 7600 indi-

C

cates communication failure.

company 〔可〕会社	機長は、会社に到着遅延を通報する。 The captain informs the company of arrival delay.
company radio 会社の無線（局）	同じ会社の航空機の空対空通信には、会社無線局の周波数が使われる。 The company radio frequency will be used for air to air communication of the same company's aircraft.
compare 〔他〕比較する	運航管理者は、燃料消費率を比較する。 The dispatcher compares fuel consumption rate.
comparison 〔不〕比較	性能の比較 performance comparison
complete 〔他〕完了する	パイロットは、チェックリストを完了した。 The pilots have completed the checklist.
completion 〔不〕完了	機長は、飛行の終了をATCに報告しなければならない。 The captain must report ATC the completion of flight.
compliance 〔不〕応じること	所定の手順に従った離陸 takeoff in compliance with the prescribed procedure
comply 〔自〕従う	社員は、会社の規則に従わなければならない。 Employees must comply with the company regulation.
compose 〔他〕構成する	GCAによる誘導は、監視レーダーによる誘導とPARによる誘導から成る。 Vectoring by GCA is composed of surveillance radar and precision approach radar.
composition 〔可〕構成	国内線の乗員編成は、シングル編成である。 The crew composition of domestic

C

operation is a single crew.

---

compulsory 〔形〕義務的な
compulsory education
義務教育

義務航空機には、ELTが搭載されている。
A compulsory aircraft is equipped with ELT.

---

concentrate
〔他〕～に集中する
concentration
〔不〕集中

パイロットは、離着陸中は機の操縦に集中する。
Pilots concentrate themselves on maneuvering aircraft during take-off and landing.

---

concept 〔可〕構想 概念

不安定なアプローチの排除は、安全運航の重要な概念である。
Elimination of un-stabilized approach is an important concept of flight safety.

---

concern 〔可〕関心事

出発前ブリーフィングにおける主要な関心事は、目的地の天候であった。
Our chief concern at the pre-flight briefing was the weather at the destination.

---

concise 〔形〕簡潔な

通報は、簡潔な表現であるべきである。
Messages should be made in concise expression.

concisely 〔副〕簡潔に

通報は、簡潔に表現されるべきである。
Messages should be concisely expressed.

---

conclude 〔他〕終える

パイロットは、低高度における天候を重視してブリーフィングを終らせた。
The pilot concluded his briefing with an emphasis on the weather at low altitude.

conclusion 〔可〕結論

発表は、結論に至った。
Presentation has come to the conclusion.

condition 〔不〕状態	前線系によって不安定な気象状態になった。 The frontal system brought unstable weather condition.
conduct〔他〕(業務を) 行う 処理する 管理する	パイロットの責任は、安全運航の実施である。 A pilot's responsibility is to conduct safe flight.
confirm 〔他〕確認する	機長は、指定高度を確認する。 The captain confirms the assigned altitude.
confirmation 〔不〕確認	復唱によって高度の確認がなされなければならない。 Confirmation of altitude must be made by read back.
conflict 〔不〕対立 矛盾 衝突	レーダーが提供する交通情報によって、交通の対立が防止される。 Radar traffic information prevents conflict of traffic.
confuse〔他〕混乱させる	通勤電車は、滅多に混乱しない。 Operation of commuter train is seldom confused.
confusion〔不〕混乱	標準操作手順に従っていれば、混乱は避けられる。 To follow standard operating procedures prevents being involved in confusion.
congest〔他〕混雑させる **congestion** 〔不〕過密 渋滞	空港周辺におけるレーダー誘導によって、渋滞は解消された。 Radar vectoring in the vicinity of airport has solved traffic congestion.
conjunction 〔不〕結合 **in conjunction with〜** 〜に関連して	速度のコントロールは、ピッチとスラストに関連する。 Speed control is in conjunction with pitch and thrust.

C

C

connect 〔他〕結合する **connection** 〔可·不〕接続 結合する部分	FBWシステムは、操縦入力と操舵面を電気的に結合する。 Fly by wire (FBW) system electrically connects control input with control surfaces.
consider 〔他〕熟考する	運航管理者は、飛行計画の高度選択についていろいろ考えた。 The dispatcher considered altitude selection for the flight plan.
consideration 〔可·不〕考慮	機長は、副操縦士の提案を考慮に入れた。 The captain took copilot's suggestion into consideration.
consist 〔自〕〜から成る	離陸許可は、使用滑走路番号と風のデータから成る。 Takeoff clearance consists of the active runway number and wind data.
consistent 〔形〕調和して	運航の首尾一貫した政策は、安全である。 Consistent policy of flight operation is safety.
contact 〔不〕観察 〔他〕連絡をとる	目視飛行で進入する approach by visual contact タワーと通信連絡をとる contact tower
contain 〔他〕含む	飛行計画には、所望の巡航高度を含むことが出来る。 The flight plan may contain desired cruise level.
contents 〔複〕内容物	コンテナーの内容物は、検査済みである。 Contents of the container have been checked.
continue 〔他〕継続する	パイロットは、深い霧の中進入を続けた。 The pilot continued approach in a

dense fog.

continuous
〔形〕連続的な

ひっきりなしの雨で、視程が下がった。
The continuous rain lowered visibility.

continuously
〔副〕連続的に

引き続く突風で、パイロットは、着陸で苦心した。
Continuously blowing gusty wind caused the pilot hard work at landing.

---

contribute
〔自·他〕貢献する

西風が強かったので、到着が早くなった。
Strong westerly wind contributed to an early arrival.

---

control
〔不〕管制 制御
〔他〕管制する 制御する
controller 管制官

GCAのファイナルコントローラーは、機を最終進入経路上に誘導する。
GCA final controller guides aircraft on the final approach path.

control column 操縦桿

操縦桿は、各パイロットのそれぞれの操縦席にある。
Each pilot has a control column at his/her control position.

control area 管制区
control zone 管制圏
controlled air space 管制空域

管制空域は、管制区と管制圏から成る。
Controlled air space consists of the control area and the control zone.

control transfer 管制移管

管制移管に伴って、通信移管が必要である。
Communication transfer is required in conjunction with control transfer.

---

controller pilot data-link communication (CPDLC)

管制官とパイロット間のデータリンク通信
data-link communication between controllers and pilots

---

conventional 〔形〕既存の

従来からある音声通信

C

	conventional voice communication
conversation 〔可·不〕会話	ATC通信には、流暢な会話より確実な会話が要求される。 ATC communication requires reliable conversation rather than fluent conversation.
convert 〔他〕転換する	アンテナは、電流を電波に転換する。 An antenna converts electric current into a radio wave.
core 〔可〕核心 芯	運輸多目的衛星（MTSAT）は、航空交通管理の核心をなす衛星である。 MTSAT is the core satellite of the air traffic management system.
correct	
〔形〕正確な	DMEは、電波により正確な距離を提供する。 DME provides correct distance by means of radio wave.
〔他〕訂正する 直す	管制官は、パイロットの復唱を訂正した。 The controller corrected the pilot's read back.
correction 〔不〕訂正	送信エラーは、所定の手順で訂正すべきである。 Correction by the given procedure should be taken for a transmission error.
correctly 〔副〕正確に	正確に計算された重量·重心位置情報が、出発前に発行される。 Correctly computed weight and balance sheet is issued before departure.
country 〔可〕国	航空機は、他国の上空を通過する。 Aircraft fly over the other countries.
course 〔可〕コース 針路	針路から外れるには、ATCの許可が必

要である。
Course deviation requires ATC approval.

coverage
〔可·不〕範囲 覆域
レーダー覆域は、周波数により変化する。
Radar coverage varies in accordance with frequencies.

cumulus 〔可·不〕積雲
積雲は、飛行に好ましくない。
Cumulus adversely affects flight.

cumulonimbus (Cb)
〔可·不〕積乱雲
パイロットは、積乱雲に近づくことを避ける。
Pilots avoid to go near the cumulonimbus.

current 〔形〕現在の
現在の気象情報が常に入手可能である。
The current weather information is always available.

〔可·不〕電流
整流器は、交流を直流に変換する。
The rectifier changes alternating current into direct current.

cutback 〔可〕削減
離陸の騒音軽減には、パワー削減が利用される。
Power cutback is applied to noise abatement at takeoff.

D

danger 〔可〕危険
制限速度を超えることは危険の要因である。
Exceeding the speed limit is a danger.

dangerous 〔形〕危険な
山岳波は、危険である。
Mountain wave is dangerous.

data 〔複〕データ
風のデータは、ATISに含まれている。
Wind data is included in ATIS.

data link データリンク
パイロットと管制官の間の衛星通信を介したデータリンク通信が、利用可能である。
Data link communication between pilots and controllers through satellite communication is available.

---

decide 〔他〕決心する
パイロットは、着陸復行を決心した。
The pilot decided to go-around.

decision 〔不〕決定 決断
100フィートは、着陸を決心する高度である。
100 feet is the decision altitude of landing.

---

decimal
〔可〕少数 〔形〕少数の
小数点　decimal point

---

declare 〔他〕宣言する
機長は、緊急時の権限の執行を宣言した。
The captain declared execution of emergency authority.

declaration 〔不〕宣言
緊急着陸の宣言
declaration of emergency landing

---

decrease 〔不〕減少 縮小
パイロットは、速度の減少を認めた。
The pilot recognized decrease in speed.

〔自･他〕
減少する 減少させる
パイロットは、速度を減少した。
The pilot decreased the speed.

---

define 〔他〕定義する
専門用語は、公文書に定義されている。
Technical terms are defined in the official document.

definition 〔可〕定義
専門用語の定義は難解である。
Definition of technical terms are the problem.

---

degree 〔可〕度
外気温は、－50度である。
The ambient temperature is －50 degrees.

---

delay
〔不〕遅延
到着遅延が発表された。

	They announced arrival delay.
〔他〕遅らせる	向かい風で到着が遅れた。 Headwind component caused arrival delay.
deliver 〔他〕伝える	交通量の多い場所では、レーダーによる交通情報が航空機に伝えられる。 Radar traffic information is delivered to aircraft in the heavy traffic area.
delivery 〔可·不〕伝達	NOTAMの伝達は、一括呼出しで行われた。 Delivery of the NOTAM was made by means of all station call.
delivery section 伝達席	伝達席は、パイロットに管制承認を中継する。 The delivery section relays flight clearance to pilots.
Delinger phenomenon デリンジャー現象	強い太陽活動によりデリンジャー現象が起こる。 Strong sun's activity causes Delinger phenomenon.
demodulate 〔他〕復調する	受信機には、受信信号を復調するための装置がある。 A receiver includes a device to demodulate received signal.
depart 〔自·他〕出発する	18番ゲートから10時15分に出発する。 The flight departs from gate 18 at 10:15.
departure 〔可·不〕出発	この時間には、多くの出発がある。 There are many departures at this time.
departure control 出域管制	出域管制は、レーダー誘導を提供する。 Departure control provides aircraft with radar vectoring.
departure frequency 出域周波数	出域周波数は、通常フライトクリアランスに含まれている。

D

	Departure frequency is normally included in the flight clearance.
depend 〔自〕～に頼る	安全運航は、人の和による。 Flight safety depends on the harmony among people.
descend 〔自〕降下する	パイロットは、すぐに降下し始めた。 The pilot quickly started to descend.
descent 〔可·不〕降下	降下の許可は、まだ出ていない。 Descent clearance is not issued yet.
describe 〔他〕記述する	標準操作手順は、規定に記述されている。 The standard operating procedures are described in the manual.
description 〔不〕記述	機長は、気象に関する手短な記述を提出した。 The captain gave a brief description of weather.
designate 〔他〕任命する	パイロットが機長に任命される。 A pilot is designated as the commander.
designated short training course 認定短期養成コース	認定短期養成コースには誰でも参加できる。 The designated short training course is open to anyone.
desire 〔他〕望む	飛行計画の所望のコースがクリアランスに与えられるであろう。 The course desired by the flight plan will be given in the flight clearance.
destination 〔可〕目的地	目的地の予報は良い。 The weather forecast at the destination is favorable.
detect 〔他〕検出する	TCASは、危険なトラフィックを検出する。 TCAS detects hazardous aircraft.

D

detect-ability 探知能力

レーダーの探知能力は、周波数により異なる。
The detect-ability of radar differs by its frequency.

detectable
〔形〕探知可能な

小さいターゲットは、高い周波数のレーダーにより探知できる。
A small target is detectable with a radar using high frequency.

detective 〔形〕探知用の
detective device
探知装置

レーザージャイロは、機の運動の探知装置である。
The lazar-gyro is a detective device of aircraft's motion.

---

determine 〔他〕決定する

パイロットは、無線標識によって位置や方位を決定する。
Pilots determine a fix or a bearing by means of radio beacon.

---

deviate 〔自〕逸脱する

指定高度から逸脱するには、ATCの許可が必要である。
ATC approval is required to deviate from the assigned altitude.

deviation 〔可･不〕逸脱

標準操作手順からの逸脱は許されない。
Deviation from the standard operating procedures is not allowed.

---

dew point 露点

温度と共に露点も報告される。
Dew point is reported together with the temperature.

---

difference 〔可・不〕相違

わずかな周波数の相違が、種々の問題になる。
A slight difference in frequency results in various troubles.

different 〔形〕異なる

パイロットは、異なる周波数で再送信を試みる。
A pilot makes another attempt to transmit on a different frequency.

differential 〔形〕差異の
differential GPS (DGPS)

MTSATは、擬似衛星航法（DGPS）の運用範囲を拡大した。

MTSAT has expanded the service area of DGPS.

---

difficult 〔形〕難しい
突風の中での速度の制御は難しい。
Speed control in gusty wind is difficult.

difficulty 〔可〕難しさ
不意の風の変化で、降下経路の処理が難しくなった。
Unexpected wind shift caused difficulties in path management of descent.

---

diffract 〔他〕回折する
地表面に沿って進んでいる電波は、回折されている。
Radio waves traveling along the surface are diffracted.

diffraction 〔不〕回折
地上波は、回折により遠くへ伝播する。
Ground surface waves reach far distance by diffraction.

---

direct

〔形〕直接の
直接呼出し direct call

〔他〕指揮する 管理する
機長は、運行を指揮する。
PIC directs flight operations.

direct current 直流 **(DC)**
直流電圧は、トランスでは調整できない。
DC voltage can't be adjusted by a transformer.

directly 〔副〕直接に
その局は、直接コンタクトするには遠すぎる。
The station is too far to directly contact.

direction
〔可〕方向
ターゲット機の方向は、2時の方向である。
The target aircraft is at 2 o'clock direction.

〔不〕指示
パイロットは、管制官の指示に従う。
Pilots comply with controller's direction.

directional
〔形〕指向性の 方位の

レーダーシステムには指向性のアンテナがある。
Radar system is equipped with a directional antenna.

---

discipline 〔不〕規律 統制

通信においては、高度の規律が重要である。
The highest standard of discipline in communication is a matter of importance.

---

discontinue 〔他〕中断する

パイロットは、離陸を中断した。
The pilot has discontinued takeoff.

discontinuance
〔不〕中断

離陸中断は、直ちに報告された。
Discontinuance of takeoff was immediately reported.

---

discretion 〔不〕裁量

航法は、パイロットの裁量の下にある。
Navigation is within pilots discretion.

---

dispatch 〔他〕派遣する
dispatcher
運航管理者の通称

ディスパッチャーの正式の名称は、運航管理者である。
The formal name of a dispatcher is a flight operations officer.

---

display 〔可·不〕表示
〔他〕表示する
display unit 表示装置

DMEの距離は、デジタル表示装置に表示される。
DME distance is displayed on a digital display unit.

---

disregard 〔他〕無視する

パイロットは、管制官の指示を無視することは出来ない。
No pilot can disregard controller's instruction.

---

distance 〔可·不〕距離

DMEの距離は、斜距離である。
DME provides slant distance.

distance line 距離線

3本の距離線の交点が、位置である。
Intersection of three distance lines is the position.

distance measuring

DMEは、質問機と応答機から成る。

equipment（DME）	DME consists of an interrogator and a transponder.
distinct〔形〕はっきりした	通報は、はっきりした表現であるべきである。 Messages should be made in a distinct expression.
distinctly 〔副〕はっきりと	通報は、はっきり発音されなければならない。 Messages must be distinctly pronounced.
distress〔不〕遭難 災難	MAYDAYは、遭難を意味し通信の沈黙を命じる。 MAYDAY means distress and commands communication silence.
〔他〕悩ます 悲しませる	そのニュースは、多くの人々を悲しませた。 The news has distressed many people.
distress call 遭難呼出し	遭難呼出しには、絶対優先権がある。 Distress calls have absolute priority.
distress condition 遭難状態	フライトが遭難状態にあるか否かは、機長が決定する。 The captain decides if the flight is in distress condition or not.
distress traffic 遭難通信	遭難通信を妨害してはならない。 No one has to interfere distress traffic.
divert〔他〕転換する	目的地空港の視程が悪く、代替空港へ行くことになった。 Low visibility at the destination airport made the flight to divert to the alternate airport.
diversion〔可〕転換	代替空港への行先変更 diversion to the alternate airport
divide〔他〕分類する	タワーは、三つのセクションに分けられている。

D

	The tower is divided into three sections.
division 〔可〕 区分	タワーには三つのセクションがある。 The tower has three divisions.
document 〔可〕書類 文書	この文書は、読み難い。 This document is hard to read.
domestic 〔形〕国内の	航空会社は、大型双発ジェット機を国内運航に採用している。 Airlines have introduced large twin jets to domestic operation.
doubt 〔可·不〕疑い	霧が晴れることに疑いは無い。 There is no doubt of the fog clears up.
〔自·他〕疑う	ディスパッチャーは、霧がすぐに晴れるとは信じていない。 The dispatcher doubts if the fog clears up soon.
drive 〔他〕駆動する	発電機は、エンジンで駆動される。 A generator is driven by an engine.
due to ～のため	視界が悪いため、滑走路は閉鎖された。 Due to the low visibility, the runway was closed.
dump 〔他〕投棄する	パイロットは、着陸重量を減らすために燃料を放棄する。 Pilots dump fuel to reduce landing gross weight.
duration 〔不〕期間	巡航中ずっと揺れていた。 It was bumpy for the duration of entire cruise.
during 〔前〕～の間	パイロットは、離陸及び着陸中は呼出しに答えない。 Pilots will not respond to a call during takeoff and landing.

E	
each 〔形〕各々の	各航空機にはSELCALコードが与えられている。 Each aircraft is assigned with a SELCAL code.
easy 〔形〕平易な	通報は、平易な表現で送信されるべきである。 Message should be transmitted with an easy expression.
economical 〔形〕経済的な **economically** 〔副〕経済的に	経済的な燃料の使用は、飛行性能を改善する。 Economical use of fuel improves flight performance.
effect 〔可〕効果 影響	ハードランディングの経験は、良い経験となった。 Hard landing experience contributed a good effect.
effective 〔形〕有効な	有効な経験　effective experience
efficient 〔形〕効率の良い	能率的な高度の選定 efficient altitude selection
efficiently 〔副〕効率良く	パイロットは、長距離飛行では、効率良くステップクライムする。 On a long range flight, pilots make step-climb efficiently.
efficiency 〔不〕効率	大型ファンジェットエンジンの燃料効率は、素晴らしい。 Fuel efficiency of large fan-jet engines is excellent.
either 〔形〕どちらでも	航空機は、どちらのパイロット席からでも操縦可能である。 Aircraft can be controlled from the pilot's seat of either side.
elapse 〔自〕経過する **elapsed time** 経過時間	アンテナと局の間の電波の経過（伝播）時間は、距離に相当する。 Radio wave's elapsed time between the antenna and the station is

	equivalent to the distance.
electric 〔形〕電気の	航空機の電力は、エンジン駆動の発電機により供給される。 Aircraft electric power is supplied by an engine driven generator.
electricity 〔不〕電気 電流	電流は、電波に転換される。 Electricity is converted into radio waves.
electro-magnetic wave 電磁波	電波は、電磁波である。 Radio waves are electro-magnetic waves.
element 〔可〕要素	ICAO標準用語は、ATC通信の主要要素である。 ICAO standard phraseology is the main element of ATC communication.
elevation 〔不〕高さ	飛行場標点は、標高を示す。 The airport reference point indicates its elevation.
eliminate 〔他〕除去する	不必要な用語は、通報から除去すべきである。 Unnecessary words should be eliminated from a message.
emergency 〔不〕緊急 有事	トランスポンダーコード7700は、機が緊急状態にあることを示す。 Transponder code 7700 indicates the aircraft is in an emergency.
〔形〕緊急の	ATCは、緊急通報を受信した場合直ちに緊急措置を開始する。 ATC will initiate emergency procedures immediately after receiving such message.
emergency frequency 非常用周波数	通常非常用周波数は、緊急呼出しには使われない。 Normally the emergency frequency is not used for an emergency call.
emergency locator	義務航空機局には、ELTが搭載されな

E

transmitter (ELT) 救命無線機	ければならない。 ELT must be equipped on a compulsory aircraft station.
emit 〔他〕放射する	電波は、アンテナから放射される。 Radio waves are emitted from an antenna.
emphasis 〔可·不〕強調 **emphasize** 〔他〕強調する	重要な通報は、繰り返し送信して強調すべきである。 Important messages should be emphasized by repeated transmission.
en-route 〔副〕途中 航路上の	ATCは、航空機に航路上の周波数を指定する。 ATC assigns aircraft en-route frequencies.
encounter 〔他〕遭遇する	(航空) 機は、晴天乱気流に遭遇した。 The aircraft encountered clear air turbulence (CAT).
endeavor 〔自〕努力する	新人パイロットは、進入段階を確実に実行しようと努力したが、速度制御は正確には出来なかった。 The new pilot endeavored to manage approach path but resulted in poor speed control.
endurance 〔不〕耐久性	航空機の持久性は、燃料消費率との相関関係である。 Aircraft endurance is a function of fuel consumption rate.
engage	
〔自〕従事する 携わる	副操縦士は、機の操縦に従事した。 The co-pilot engaged himself in maneuvering the aircraft.
〔他〕従事させる (興味などを) 引く	パイロットは、改定された風のデータに気を取られた。 The pilot's attention was engaged in the revised wind data.

E

〔形〕従事している

(be engaged in ～) 彼は、通信業務に従事している。
He is engaged in a communication service.

---

engine 〔可〕エンジン
engine driven
エンジン駆動の

ジェットエンジンは、強力である。
A jet engine is powerful.

---

ensure 〔他〕確保する

通信は、確保されている。
Communication has been ensured.

---

entire 〔形〕全体の

乗客は、飛行中ずっと快適を楽しんだ。
Passengers enjoyed comfortableness for the entire flight.

---

enunciate
〔自〕（はっきり）発音する
enunciation 〔不〕発音

彼は、言葉をはっきり発音する。
He enunciates his words.
彼の発音は、はっきりしている。
His enunciation is clear.

---

environment
〔可·不〕環境
environmental
〔形〕環境の
environmental pollution
環境汚染

レーダーの環境は、厳しい。
Radar environment is severe.
航空機による環境汚染は、研究されている。
Environmental pollution by aircraft is under study.

---

equal
〔形〕等しい 平等な
〔他〕匹敵する

無線周波数の使用に当たって、我々は皆平等の権利を有する。
All of us have equal rights in using radio frequencies.

equalize
〔他〕等しくする

イクオライザーは、周波数のずれを制御する。
The equalizer controls stray frequencies.

---

equator 〔単〕赤道

MTSATは、赤道上空に位置する。
MTSAT is located over the equator.

---

equip 〔他〕設備する

多くの滑走路に、ILSが装備されている。

E

	Many runways are equipped with ILS.
equipment 〔不〕機器	ILSの機上装備 ILS airborne equipment
equitable 〔形〕公平な **equitably** 〔副〕公平に	無線周波数は、各国で公平に使われている。 Radio frequencies are equitably used by every country.
equivalent 〔形〕同等の	電波がアンテナとターゲットの間の往復に要した経過時間は、距離に等しい。 The elapsed time spent by a radio wave for a round trip between the antenna and the target station is equivalent to the distance.
error 〔可･不〕誤り **erroneous** 〔形〕間違った **erroneously** 〔副〕間違って	誤りをする　make an error 間違った通報は、直ちに訂正されなければならない。 An erroneous message must be immediately corrected.
essential 〔形〕必須の	運航においては、機敏な行動が欠かせない。 To take prompt action is essential in flight operation.
establish 〔他〕設置する 設立する 設定する	無線局は、他の局の通信を妨げないように設置されている。 A radio station is so established that it does not interfere communication of other stations.
establishment 〔不〕設立　設定	パイロットは、通信設定を試みた。 The pilot tried to make communication establishment.
estimate 〔可〕推定	予定到着時刻 estimated time of arrival (ETA)
〔他〕推定する	風は、270度、25ノットと推定されていた。

E

Wind was estimated to be 270/25.

every 〔形〕すべての

皆が、快適な旅を願った。
Every person hoped for comfortable journey.

evidence 〔不〕証拠

検査の結果、回路に過電流が流れたことが判明した。
Inspection revealed evidence of excessive current in the circuit.

evident 〔形〕明らかな
evidently 〔副〕明らかに

明らかにコースから外れていた。
It is evident that they were off course.

exact 〔形〕正確な

ICAO標準用語は、各語の正確な意味で使われなければならない。
ICAO standard phraseology must be used in the exact meaning of each word.

exactly 〔副〕正確に

各語の意味は、厳密に指定されている。
The meaning of each word is exactly specified.

examine 〔他〕検査する

過去の整備記録が検査された。
The past maintenance records have been examined.

examination 〔可〕試験
exam （短縮形）

明日試験がある。
We'll have an exam tomorrow.

example 〔可〕例

オートパイロットは、操縦の良い手本を示す。
The autopilot gives us a good example of aircraft maneuvering.

exceed 〔他〕超過する

運用限界を超えることは、危険である。
It is critical to exceed operating limitations.

except
〔前〕～を除いては
〔他〕除外する

出発地から9キロメートル以内のVFR機を除くすべてのフライトは、飛行計

E

	画をファイルしなければならない。 All flights except VFR within 9 km of the departing point are required to file a flight plan.
exception 〔可〕例外	他に除外例は無い。 There is no other exception.
exceptional 〔形〕例外的な	この大雪は、この空港にとっては例外的なことである。 This heavy snow is exceptional for this airport.

exchange 〔可·不〕やり取り 交換	通信のトラフィックとは、通報の交換のことである。 Communication traffic is an exchange of messages.

execute 〔他〕実行する	燃料投棄を実施する execute fuel dumping
execution 〔不〕実行	機長は、緊急時職権の実行を発表した。 The captain has announced execution of an emergency authority.

exempt 〔他〕免除する **exemption** 〔不〕免除	一括呼出しに対する受信証は、免除されている。 Acknowledgement of receipt for all station call is exempted.

exercise 〔可〕運動 練習 〔他〕運動する 練習する	ICAOのアルファベット文字の発音練習が推薦される。 An exercise in articulation of ICAO alphabetic letters is recommended.

exist 〔自〕存在する	SELCALは、選択呼出しを可能にするために現存の受信機に加えられている。 SELCAL is added to the existing receiver to enable selective call.
existence 〔不〕存在	危険な航空機の存在が、TAにより警告された。 Existence of a hazardous aircraft has been warned by traffic advi-

sory (TA).

expand

〔自〕拡大する	フラップを出すと、翼面積が広がる。 Wing area expands by flap extension.
〔他〕拡大する	燃料消費率の向上は、航続時間を拡大する。 Improvement in fuel consumption rate expands endurance.
expansion 〔不〕拡大	RNAV空域の拡大は、管制を容易にする。 Expansion of RNAV area facilitates traffic control.

expect 〔他〕予期する	パイロットは、進入開始時までには天候が回復することを予期していた。 Pilots expected weather recovery by the time of approach initiation.
expectation 〔不〕予期	予期したとおり天候は回復した。 The weather recovered according to expectation.
expected approach time (EAT) 進入予定時刻	管制官は、SA101にEATを知らせた。 The controller informed SA101 of its EAT.

expedite 〔他〕促進する	管制官は、パイロットに降下開始を急ぐように指示した。 The controller instructed the pilot to expedite to start descent.

express 〔他〕表現する	通報は、簡潔に表現されるべきである。 Messages should be concisely expressed.
expression 〔可·不〕表現	緊急状態の表現方法は、ICAOの通信手順に記載されている。 Expression method of emergency condition is described in ICAO communication procedures.

extend 〔他〕拡張する	衛星は、電力供給のために太陽電池を

E

	拡げる。 Satellites extend solar cells to supply power.
extension 〔不〕拡張	降着装置を出すことにより抗力が増加する。 Landing gear extension causes drag increase.
external 〔形〕外部の	無線従事者は、無線機器の外部調節を行うことが出来る。 External adjustment of radio equipment may be made by a radio operator.

F

facilitate 〔他〕容易にする	事前の情報により、飛行の実施が容易になった。 Advance information facilitated flight management.
facilities 〔複〕施設	国内線航路には、ハイテク航法施設が装備されている。 The domestic routes are equipped with high technical navigation facilities.
factor 〔可〕要因	蒸気が、火災警報作動の要因であった。 Vapor was the factor of fire warning activation.
fade 〔自〕次第に薄れる	太陽の活動により、電波は次第に弱くなる。 Sun's activities cause radio waves to fade out.
fail 〔自〕失敗する	パイロットは、地上局との連絡に失敗した。 The pilot failed to contact a ground station.
failure 〔不〕失敗	最終進入時の速度制御が正確に出来ていないと、不安定な着陸となる。 Failure in accurate speed control

during final approach results in insecure landing.

---

far 〔副〕遠くへ

電波は、遠くへ伝播する。

Radio waves propagate far distance.

---

feature 〔可〕特徴

FMシステムの特徴は、清澄な音色である。

The feature of FM system is clarity of a tone.

---

feed

〔他〕与える 供給する

送信機は、変調搬送波を送信アンテナに与える。

A transmitter feeds the modulated carrier wave to the transmitting antenna.

feeder 〔可〕 フィーダー 給電線

同軸ケーブルは、広く用いられている給電線である。

Coaxial cable is a widely used feeder.

---

file 〔他〕ファイルする 提出する

IFR運航のオペレーターは、飛行計画をファイルする。

An operator of IFR flight files the flight plan.

---

final 〔形〕最終の

ATCの最終目的は、航空機の衝突防止である。

The final object of ATC is to prevent collision between aircraft.

final approach 最終進入

最終進入は、タワーが取り扱う。

Tower controls the final approach segment.

final approach course 最終進入コース

最終進入コースは、滑走路中心線の延長である。

The final approach course is an extension of the runway center line.

finally 〔副〕最終的に

パイロットは、ホールディングの後最終的に無事着陸した。

The pilot finally landed safely after holding.

---

find 〔自·他〕見つけ出す

ADFは、NDB局への方位を見つけ出すために使われる。

ADF is used to find the direction to an NDB station.

---

fire 〔可·不〕火 火災

火災の発生は、煙探知機で検出される。

Occurance of fire is detected by a smoke detector.

fire warning 火災警報

火災警報の発生

activation of fire warning

fire extinguisher 消火器

操縦室にも客室にも消火器が搭載されている。

Fire extinguishers are installed both in the cockpit and the passenger compartment.

---

firm 〔形〕しっかりした

パイロットは、しっかりした着陸を行った。

The pilot made firm touch down.

firmly 〔副〕しっかりと

機は、しっかりと接地した。

The aircraft firmly touched down.

---

fix 〔可〕位置

次のポジションは、北緯35度、東経135度である。

Next fix is N35.00, E135.00.

〔他〕固定する

整備員は、受信機を無線機棚に固定した。

The maintenance person fixed the receiver on the radio rack.

fixed service 固定業務

固定業務は、地上局により取り扱われる。

Fixed service is handled by ground stations.

---

flare 〔可·不〕フレアー

パイロットは、フレアーするために操縦桿を引いた。

The pilot pulled the control column for flare.

---

flight 〔可·不〕飛行

この天候での飛行は、快適に違いない。

Flight in this weather must be comfortable.

flight clearance
フライトクリアランス
管制承認

フライトクリアランスは、発行されている。

The flight clearance has been issued.

flight identification
飛行便名

飛行便名は、有償飛行の航空機の呼出符号に採用されている。

The flight identification is employed in the aircraft call sign of revenue flight.

flight information region（**FIR**）飛行情報区

日本は、福岡FIRにおいて所定の業務を提供する責任がある。

Japan is responsible to provide specified service in Fukuoka FIR.

flight level
高度 フライトレベル

フライトレベルは、QNEを基準とする14,000フィート以上における気圧高度である。

Flight level is a pressure altitude based on QNE and above 14,000 feet.

flight operation 運航

安全運航は、最善の乗客へのサービスである。

Safe flight operation is the best passenger service.

flight phase 飛行の段階

各飛行段階ごとに制限速度は異なる。

Speed limit is different for each flight phase.

flight progress
飛行の進行（経過）

パイロットは、飛行経過に伴う残存燃料を気にする。

Pilots are concerned about the remaining fuel in relation to flight progress.

flight safety 飛行の安全

飛行の安全は、最優先の運航の概念である。

Flight safety is the highest priority concept of operation.

---

flow
〔単〕流れ

交通の流れは、レーダーでコントロー

	ルされる。 Traffic flow is controlled by radar service.
〔自〕流れる	燃料は、フィルターを介してFCUにスムーズに流れて行く。 Fuel smoothly flows into the fuel control unit (FCU) through a filter.
follow 〔自·他〕続く 従う	小型機は、飛行場が見えるまで旅客機に続いて行った。 The small aircraft followed the airliner until the pilot visually contacted the airport.
foresee 〔他〕予知する	パイロットは、風の変化を予知していた。 The pilot has foreseen a wind shift.
form	
〔不〕形 型 方式 書式	飛行計画は、所定の型に記入される。 Flight plan is entered in the fixed form.
〔他〕形成する	VOR/DME局を結んで航空路を形成する。 A chain of VOR/DME stations forms an airway.
frequency 〔可〕周波数	電波の周波数は、音波の周波数より高い。 Radio waves' frequencies are higher than the sound waves'.
frequency band 周波数帯	電波の最も高い周波数帯は、ミリ波帯である。 The highest frequency band of radio waves is the extreme high frequencies (EHF).
frequency modulation (FM) 周波数変調	FM方式は、AM方式に比べ有利な点が多い。 FM system has many advantages over the AM system.
frequent 〔形〕頻繁な	この飛行場では、頻繁な大雪のため通

	常の運用が阻害される。 A frequent heavy snow obstructs normal operation at this airport.
frequently 〔副〕 頻繁に	ここには大雪が頻繁にやってくる。 A heavy snow comes here frequently.
fuel 〔可·不〕 燃料 **fuel consumption** 燃料消費	航空機は、多大な燃料を消費する。 An aircraft consumes a large quantity of fuel.
fuel dumping 燃料投棄	燃料投棄は、非正常操作の一つである。 Fuel dumping is one of abnormal procedures.
full 〔形〕 全部の	停止に滑走路の全長が必要であった。 Full length of the runway was used to stop.
fully 〔副〕 十分に	パイロットは、どんな状態においても機の取扱いに有能である。 Pilots are fully capable of aircraft handling in any condition.
function 〔可〕 機能 関数	水平安定板の機能は、機の縦安定を確保することである。 The function of the horizontal stabilizer is to secure longitudinal stability of an aircraft.
functional 〔形〕 機能上の	装置は、機能上の要件を満たしている。 The device meets the functional requirement.
malfunction 〔可〕 故障 (**mal** は、「不完全な」の意)	検査官は、故障記録を調べた。 The inspector examined the malfunction record.
further 〔副〕 より遠く	目的地よりさらに遠くへ行く燃料は無い。 The remaining fuel is insufficient to go further than the destination.

F

G	
gear 〔可〕（着陸）装置	着陸装置は、抵抗を増やすことに有効である。 The landing gears are effective to increase drag.
general 〔形〕全般的な	全般的理解は、勉強の入り口である。 General understanding is an entry of study.
general call 総括呼出し	地上局は、適当と思われる時には総括呼出しを行う。 A ground station makes general call when it is appropriate.
generate 〔他〕発生させる **generator** 発電機	発電機は、電気を発生させる。 A generator generates electricity.
generic 〔形〕包括的な **generic term** 包括的用語	エマージェンシーは、遭難と緊急の両者を意味する包括的な用語である。 Emergency is a generic term meaning distress and urgency.
glide slope 最終進入角 **glide path** 最終進入角（電波）	機は、正確にグライドスロープ上を進入した。 The aircraft correctly followed the glide slope.
global positioning system（GPS） 衛星航法システム	GPSの航法精度は、MTSATにより改善された。 GPS navigation accuracy has been improved by MTSAT.
go ahead 先へ進める どうぞ	“Go ahead”は、相手局の通信を促す。 “Go ahead” presses the other station for communication.
go-around 着陸復行	パイロットは、突風のため着陸復行した。 The pilot executed go-around because of a gust of wind.
grant 〔他〕承諾する	管制官は、高度変更要求を承諾した。 The controller granted the request

G

of an altitude change.

gross 〔形〕総体の

gross weight 総重量

機体の総重量は、最大着陸重量に近かった。

The aircraft gross weight was close to the maximum landing weight.

ground

〔可〕地面 地表

電波は、地面に沿って回折される。

Radio waves are diffracted along the ground.

〔不〕海底 アース

燃料ポンプのアースは、確実に留められていなければならない。

The ground wire of the fuel pump must be securely fastened.

〔形〕地上の

地表波は、可視線より少し遠くまで届く。

Ground surface waves reach a little further than the line of sight.

ground control 地上管制

地上管制は、走行区域の交通を監視する。

The ground control monitors traffic on the maneuvering area.

ground controlled approach (GCA) 地上誘導着陸方式

航空会社のパイロットは、GCAよりもILSを好む。

Airline pilots prefer ILS approach than GCA.

ground personnel 地上員

航空機のプッシュバックは、地上員の責任で行われる。

Ground personnel are responsible for pushing back the aircraft.

ground speed (GS) 対地速度

IASに風の成分による修正を加えた速度が、対地速度である。

IAS corrected with wind component becomes the ground speed.

ground station 地上局

移動業務は、地上局と航空機局の間の通信である。

Mobile service is the communication between ground stations and aircraft stations.

group 〔可〕集団	鳥の群れが、飛行機の離着陸を遮る。 A group of birds interrupts takeoff and landing of aircraft.
guide 〔他〕誘導する	管制官は、航空機を安全に誘導する。 A controller guides an aircraft safely.
guidance 〔不〕誘導	パイロットは、管制官の誘導に従う。 A pilot follows the guidance of a controller.
gust 〔可〕突風	機は、突風によって衝撃を受けた。 The aircraft was struck by a gust.
gusty 〔形〕突風性の	強い突風性の風があった。 There was a strong gusty wind.

H

habit 〔可·不〕習慣	言語習慣により発音が異なる。 Language habit makes difference in pronunciation.
handle 〔他〕取り扱う	貨物の取り扱いに時間がかかっている。 Cargo handling is taking time.
hard 〔形〕激しい 過酷な	激しい雨のため、着陸復行した。 A hard rain caused execution of go-around.
harm 〔不〕〔他〕害 損害 **harmful** 〔形〕有害な	不必要な会話は、有害な混信である。 Exchange of unnecessary conversation is harmful interference.
hazard 〔可〕危険 **hazardous** 〔形〕危険な	機の周辺の危険範囲内の交通は、TCASで監視されている。 Traffic in the hazardous area around the aircraft is monitored by TCAS.
heading 〔可〕機首方位	機首方位270° を維持せよ。 Keep the heading of 270° .
hereby 〔副〕これによって	〔これにより～を宣する〕など儀式·

H

	法律文書などに使われる。
high frequency 高周波	高周波の電波は、電離層で反射される。 High frequency radio waves are reflected by the ionosphere.
hold 〔自･他〕保持する **holding** 〔不〕待機	NDB局上空での待機は、許可された。 Holding over NDB station has been approved.
however 〔副〕しかしながら	嵐は過ぎたが、小雨は引き続き降っていた。 The storm has passed, however, a light rain continued.

I

icing 〔不〕着氷	機体は、着氷から守られている。 The airframe is protected from icing.
identify 〔他〕確認する	パイロットは、滑走路を目視確認した。 The pilot identified the runway in sight.
identification 〔可･不〕確認 識別	航空機の識別は、コールサインによって行われる。 Aircraft identification is made by its call sign.
if 〔接〕もし～ならば **if A or B** **A**であるか**B**であるか	遭難状態であるか緊急状態であるかは、機長が決定する。 PIC determines if it is distress condition or urgency condition.
immediate 〔形〕即時の	遭難状態は、即時の援助を必要とする。 Distress condition requires immediate assistance.
immediately 〔副〕直ちに	ATCは、遭難機のことを直ちにRCCに知らせた。 ATC immediately informed RCC of

I

the aircraft in distress condition.

imminent 〔形〕切迫した	切迫した危険が予知される。 Imminent danger is foreseen.
impedance 〔不･単〕 インピーダンス **Impedance matching** インピーダンス整合	送信機とアンテナ間のインピーダンス整合は、フィーダーによってとられている。 Impedance matching between the transmitter and the antenna is made by a feeder.
importance 〔不〕重要性	パイロットは、標準操作手順の重要性について熟知しているべきである。 Pilots should be well familiar with the importance of standard operating procedures.
important 〔形〕重要な	ある周波数で送信する前に聴守することは、重要なことである。 Listening on a frequency before transmitting is important.
impose 〔他〕賦課する 課する **imposition** 〔不〕賦課すること	MAYDAYは、通信の沈黙を課する。 MAYDAY imposes communication silence.
impossible 〔形〕不可能な	この重量でFL370を巡航することは出来ない。 Cruising at FL 370 is impossible with this weight.
inbound 〔形〕帰航の	帰航便が混雑している。 Inbound traffic is congested.
include 〔他〕含む	搭載燃料には、予備燃料が含まれる。 Reserve fuel is included in the fuel to be loaded.
increase 〔不〕増加	向かい風成分の増加は、好ましくない。 Increase in headwind component is an unfavorable matter.

〔自·他〕増加する	冬季には、西行便の到着遅れが増える。 Arrival delay of west bound flights increases in winter season.
indicate 〔他〕表示する	航空機の高度計は、気圧高度をフィートで表示する。 The aircraft altimeter indicates pressure altitude in feet.
indication 〔可·不〕表示	PAPIは、精密目視グライドスロープ表示を提供する。 PAPI provides precision visual glide slope indication.
indicator 〔可〕表示器	DMEの表示は、デジタル表示器である。 DME display is a digital indicator.
individual 〔形〕個々の	アンテナ素子の長さは、個々の周波数帯によって異なる。 Antenna element length differs by individual frequency band.
individually 〔副〕個々に	管制官は、SSRを使って特定の航空機を個別的に識別する。 The controller individually identifies a specific aircraft by use of SSR.
ineffective 〔形〕無効の 効果的でない	VHFは、長距離通信には効果的でない。 VHF is ineffective for long range communication.
inertial navigation system (INS) 慣性航法装置	INSは、航法システムの核である。 INS is the core of navigation system.
Inertial reference system (IRS) 慣性基準装置	IRSは、飛行制御システムの核である。 IRS is the core of flight management system.
inevitable 〔形〕必然的な	短い滑走路で最大離陸パワーを使用することは、必然的である。 Use of the maximum takeoff

I

thrust for a short runway is inevitable.

inform 〔他〕通知する	パイロットは、受信機の不具合を整備に通知した。 The pilot informed the maintenance section of a malfunctioning receiver.
information 〔不〕情報	重要な気象情報は、航空機に速やかに伝達される。 Significant weather information is promptly delivered to the aircraft.
initial 〔形〕最初の	通信設定における最初の言葉は、受信局の呼出符号である。 The initial word in communication establishment is the call sign of the station being addressed.
initiate 〔他〕開始する	管制官は、レーダー誘導を開始した。 The controller has initiated radar vectoring.
initiation 〔不〕開始	降下開始が遅れた。 Descent initiation was late.
inspect 〔他〕点検する	ILS電波の精度は、定期的に検査される。 Accuracy of ILS radio waves is periodically inspected.
inspection 〔可·不〕検査	検査の結果、システムは正常であることが判明した。 Inspection found the system is normal.
install 〔他〕装置する **installation** 〔可·不〕装置	DMEは、VORの装置に装備されている。 DME is installed on the VOR installation.
instruct 〔他〕指図する	管制官は、パイロットに機首方位を270°に変更するように指図した。 The controller instructed the pilot to change heading to 270°.

I

instruction 〔可·不〕指図	管制官の指図は、クリアランスとみなされる。 The controller's instruction is regarded as a clearance.
instrument 〔可〕計器	燃料計器は、ポンド目盛りである。 An instrument that tells fuel quantity is calibrated in pounds.
instrument approach 計器進入	ADFアプローチは、計器進入方式の一つである。 ADF approach is one of the instrument approach procedures.
instrument flight rule (**IFR**) 計器飛行方式	定期旅客機は、計器飛行方式で運航される。 Airliners are operated under IFR.
instrument landing system (**ILS**) 計器着陸方式	計器着陸方式は、進入時の安全運航に寄与する。 ILS contributes flight safety during approach.
intend 〔他〕意図する	パイロットは、進入時に最小燃料消費方式を使うように考えた。 The pilot intended to use the minimum fuel consumption method for approach.
intention 〔不〕意図	彼の意図するところは、天候のため変更された。 His intention was changed because of the weather.
intentional 〔形〕故意の **intentionally** 〔副〕故意に	接地点の故意の延長は、危険である。 Intentional extension of the touchdown point is a danger.
intercept 〔他〕 途中で遭遇する 傍受する	その遭難通報は、多くの局により傍受された。 The distress message was intercepted by many stations.
interest 〔可·不〕関心	パイロットは、火山活動に大きな関心を示した。

I

The pilot took a great interest in volcanic activity.

---

interfere
〔自〕接触する 干渉する

右翼端が、誘導路灯柱に接触した。
The right wing tip interfered with a taxiway light post.

interference
〔不〕干渉 妨害 混信

他の局による通信への干渉は禁止されている。
Interference in communication by other station is prohibited.

---

international〔形〕国際的な

internationally
〔副〕国際的に

英語は、世界的に運航に利用されている。
English is internationally applied to flight operation.

international airport 国際空港

International Civil Aviation Organization（**ICAO**） 国際民間航空機関

International Mobile Satellite Organization（**IMSO**） 国際移動通信衛星機構

International Telecommunication Union（**ITU**） 国際電気通信連合

International Standard Atmosphere（**ISA**）
国際標準大気

QNE は、国際標準大気を基準とする。
QNE is based on ISA.

---

interpilot
パイロット相互間の

同じ会社のパイロット相互間通信は、会社の無線周波数で行われるべきである。
Interpilot communication of the same company is to be made on the company radio frequency.

---

interrogate〔他〕質問する

SSRは、航空機にその識別符号を質問する。
SSR interrogates an aircraft its identification code.

interrogator
〔可〕質問機

SSRは、質問機である。
SSR is an interrogator.

I

interrupt 〔自·他〕 さえぎる 中断する **interruption** 〔可·不〕中断	騒音で通報の受信が中断された。 A noise interrupted reception of the message. レーダーサービスは、中断することなく続いている。 Radar service is continued without interruption.
interval 〔可〕間隔 合間	パルス送信の間の間隔は、反射波を受信する期間である。 An interval between pulse transmissions is a period to receive reflected signal.
intervene 〔自〕介入する **intervention** 〔可·不〕介在	管制官は、エンルート周波数で行われているパイロット間の通信に介入することがある。 A controller may intervene interpilot traffic on the en-route frequency.
introduce 〔他〕導入する **introduction** 〔可·不〕導入	輸送機に対するジェットエンジンの導入は、速度と輸送量の倍加に貢献した。 Introduction of jet engines to transport airplanes contributed to double their speed and capacity.
invitation 〔可·不〕誘い **invite**〔他〕誘う 招待する	"Go ahead" は、通信継続への誘いの用語である。 "Go ahead" is a term of invitation to continue communication.
ionosphere 〔不〕電離層	電離層は、地球大気の一部で電波を反射する。 The ionosphere is the part of the earth's atmosphere that reflects radio waves.
irrespective 〔前〕 〜にかかわりなく	この講習は、年齢に関係なく出願者に公開されている。 The training course is open to applicants irrespective of age.

issue 〔可・不〕問題点	報告には検討課題が含まれている。 The report includes an issue to review.
〔他〕発行する	飛行解析報告が、発行される。 The flight analysis report will be issued.
item 〔可〕項目	気象予報には、気になる項目は何も無い。 The weather forecast has no item to be concerned.

J

jettison 〔他〕投棄する	燃料投棄系統を使って、燃料を放棄する。 Fuel jettisoning system is used to dump fuel.
just 〔形〕ちょうど	着陸直前に雨が降り出した。 It started to rain just before landing.

K

knot 〔可〕ノット 結び目	1ノットは、時速約1,852メートルの速度である。 One knot is a speed of approximately 1,852 m/h.
know 〔自･他〕 知る 知っている	パイロットは、減速のために機の抵抗を増加する方法を知っている。 Pilots know how to increase the drag of an aircraft for deceleration.

L

land 〔可･不〕陸地 〔他〕着陸する	機は、定時に着陸した。 The aircraft landed on schedule.
landing 〔不〕着陸	パイロットは、不時着を決意した。 The pilot decided to execute forced landing.
landing roll 着陸滑走	機は、着陸滑走中に横風を受けた。 The aircraft encountered a cross-

wind during landing roll.

landing runway 着陸滑走路

着陸滑走路の一部に雪が残っていた。
The snow remained on some part of the landing runway.

---

language 〔不〕言語

ATC言語は、基本的には英語である。
Basically, the ATC language is English.

language habit 言語習慣

言語習慣による発音の相違は、標準化により是正される。
Difference in pronunciation by language habit is corrected by standardization.

---

last 〔形〕先程の 最後の

先程のクリアランスは、次のように訂正される。
The last clearance is corrected as follows.

---

lateral 〔形〕側面の

ILSの横方向の誘導電波は、ローカライザーである。
The lateral guidance signal of ILS is the localizer.

laterally 〔副〕横に 横から

機は、横に滑った。
The aircraft slipped laterally.

---

latest 〔形〕最新の

無線航法設備には、最新の技術が装備されている。
Radio navigation facilities are equipped with the latest techniques.

---

law 〔可·不〕法律 法規
lawful 〔形〕合法的な
unlawful 〔形〕非合法的な

会社の運航規定の条項は、該当する法律に基づいている。
The articles of company's operations manual are based on applicable laws.

---

leave

〔自〕出発する

出かけるべき時間です。
It's time to leave.

〔他〕出発する

その便は、東京を出てパリへ行く。

The flight leaves Tokyo for Paris.

---

level
〔可·不〕水平面 高度
フライトレベルは、大気の圧力面である。
Flight level is a surface of the atmospheric pressure.

---

license
〔可〕免許 認可
〔他〕認可する
パイロットは、無線従事者の免許を持っている。
A pilot has a radio operator's license.

---

likely 〔形〕ありそうな
雨が降り出しそうである。
It is likely to start raining.

---

limit
〔可〕制限 限度
限度を超過した速度での運航は、報告されなければならない。
Operation with a speed exceeding the limit must be reported.
〔他〕制限する
進入空域の10,000フィート以下では、速度は250ノットに制限されている。
The speed below 10,000 feet in the approach area is limited to 250 knot.

---

line 〔可〕線 路線
国内路線　domestic lines
line of sight 目視線
通常VHFの伝播範囲は、目視線の範囲である。
Normally, the VHF propagation range is within the line of sight.

---

link 〔他〕結合する
不注意とトラブルは、密接に関係している。
Lack of attention and troubles are closely linked.

---

listen
〔自〕聞く 傍聴する
無線局は、遭難通信を引き続き傍聴すべきである。
Stations should continue listening to the distress traffic.

L

local 〔形〕地方の 場所の

local time 地方時

ラジオダクトは、温度逆転のある場所の気象状態により形成される。

Radio duct is formed by a local weather condition where temperature inversion exists.

localizer ローカライザー

ローカライザーは、ILSの平面コースの誘導電波である。

Localizer is the lateral guidance radio signal of ILS.

locate

〔他〕設置する 位置する

最終進入開始点には、マーカービーコンが設置されている。

A marker beacon is located at the initial point of final approach.

location 〔可〕位置

ILSのマーカービーコンは、最終進入経路上の位置情報を提供する。

Marker beacons in ILS provide location information on the final approach course.

long 〔形〕長い

long message 長文の通報

長文の通報は、語句の間に切れ間をおいて読むべきである。

A long message should be read with pauses between phrases.

long range 長距離

HFは、長距離通信に使われる。

HF is used for long range communication.

lose 〔他〕失う

パイロットは、激しい雨のため接地直前に一瞬視界を失った。

The pilot momentarily lost visibility due to pouring rain just before touchdown.

lost communication 通信不能

トランスポンダーコード7600は、通信不能を意味する。

Transponder code 7600 means “lost communication”.

low 〔形〕低い

進入経路がグライドスロープより低い。

The approach path is lower than the glide slope.

low frequency (LF) 長波
NDBはLFを使う。
NDB operates on LF.

low level windshear
低高度突風
レーダーによる低高度突風探知の研究は、進展している。
Study of detecting low level windshear by radar is progressing.

Low range radio altimeter (LRRA)
低高度電波高度計
低高度電波高度計は、2,500フィートまで有効である。
LRRA is effective up to 2,500 feet.

M

Mach 〔不〕マッハ
多くの定期旅客機は、M.84で飛行する。
Most airliners fly at Mach 0.84.

magnetic storm
〔可〕磁気嵐
磁気嵐からの回復には、数日かかる。
Recovery from the magnetic storm takes a few days.

main 〔形〕主要な
主要滑走路の除雪は、終了している。
Snow removal of the main runway has been completed.

maintain 〔他〕維持する
パイロットは、指定高度を維持した。
The pilot maintained the assigned altitude.

malfunction 〔可〕故障
malは、（不完全な）の意
検査官は、故障記録を調べた。
The inspector examined the malfunction record.

M

maneuver
〔可〕操縦
〔他〕操縦する
操縦と通信が重なると、パイロットは忙しくなる。
Overlap of maneuver and communication makes pilots busy.

maneuvering area
走行区域
走行区域の交通は、グラウンドが管制する。
Traffic in the maneuvering area is controlled by the ground section.

manner 〔可〕方法	AM、FMあるいはPMは、変調の一つの方法である。 AM, FM or PM is a manner of modulation.
many 〔形〕多数の	ICAO標準用語には、多くの簡略表現が含まれている。 ICAO standard phraseology includes many abbreviated expressions.
marine 〔形〕海の 船舶の	海洋汚染は、重要な研究課題である。 Marine pollution is an important subject to study.
maritime 〔形〕海（事）の	気候に対する海の影響は大きい。 Maritime influence on climate is great.
mark 〔可〕マーク 〔他〕マークをつける	航空機には登録記号が付されている。 An aircraft is marked with the registration mark.
match 〔自·他〕 調和する 調和させる	アンテナは、その固有周波数と与えられた周波数がマッチしたときに共鳴する。 An antenna resonates with a given frequency when it matches the natural frequency of the antenna.
mean 〔他〕意味する	PAN PAN MEDICAL は、医療措置の必要な緊急状態を意味する。 PAN PAN MEDICAL means an urgency condition with necessity of medical service.
meaning 〔不〕意味 〔形〕意味深長な	気象報告の意味するところでは、着陸に支障は無い。 The meaning of the weather report is that there is no problem to land.
means 〔可〕方法 手段	乱気流の時にシートベルトサインをONにすることは、乗客を怪我から守

るための手段である。

Turning the seat belt sign ON in turbulent air is a means to protect passengers from injury.

---

measure 〔可·不〕寸法
〔自·他〕測定する
measurement 〔不〕測定

視程及びRVRの測定は、メートル法で行われる。

Metric system of measurement is applied to measure visibility and RVR.

---

medical 〔形〕医療の

救急箱以外の医療装備は、機上に搭載されていない。

Medical equipment except the first aid kit is not on board the aircraft.

---

meet 〔自·他〕合う 満たす

気象状態は、離着陸の要件を満たしている。

Weather condition meets the requirement of takeoff and landing minima.

---

member state
〔可〕加盟国

主要航空会社は、ICAOの加盟国の航空会社である。

Major airlines are the airline of member states of ICAO.

---

message 〔可〕通報

通報は、ICAO標準手順に従ってはっきりと発音されなければならない。

Messages should be pronounced clearly in accordance with ICAO standard procedures.

---

meteorological
〔形〕気象の

MTSATは、フライトミッションの他にさらに気象ミッション機能も持っている。

MTSAT has a function of the meteorological mission as well as the flight mission.

---

method 〔可·不〕方法

騒音軽減方式は、空港における航空機の騒音を減少する方法である。

The noise abatement procedure is

a method to reduce aircraft noise at the airport.

microwave 〔可〕 マイクロ波 極超短波

SHFとEHFは、通常マイクロ波と呼ばれている。
SHF and EHF are usually called as microwaves.

middle frequency (MF) 〔可〕中波

MFは、NDB及びラジオ放送にも適用されている。
MF is applied to NDB and radio broadcasting as well.

midpoint 〔可〕中間点

中間点は、必ずしもETPである訳ではない。
The midpoint is not necessarily to be the ETP.

minimum
〔可〕最小限度
〔形〕最小の

パイロットは、最小燃料で飛行しようと努める。
Pilots make efforts to fly with minimum fuel consumption.

minimum fuel 燃料欠乏状態

燃料欠乏は、緊急事態である。
Minimum fuel is an emergency condition.

minister 〔可〕大臣

総理大臣は、内閣の最高責任者である。
The Prime Minister is a person in charge of the Cabinet.

minute 〔可〕分

プロペラの回転は、例えば1,800 RPMのように分単位で表現される。
Propeller revolution is expressed in "revolution per minute (RPM)" such as 1,800 RPM.

miss 〔他〕～し損なう

パイロットは、通報の一部を受信し損なった。
The pilot missed to receive a part of the message.

missed approach 進入復行

悪天のためパイロットは、進入復行した。

M

The adverse weather made the pilot to execute missed approach.

mode 〔可〕モード 形態

受信状態は、PTTスイッチを押すことにより直ちに送信モードに切り替わる。

Receiving mode is quickly changed to transmitting mode by pressing the PTT switch.

moderate 〔形〕中程度の

中程度の揺れが続いた。

Moderate turbulence continued.

modulate 〔他〕変調する

搬送波は、信号波により変調される。

The carrier wave is modulated by a signal wave.

modulation 〔不〕変調

modulator 変調器

送信機における特徴的な機器の一つは、変調器である。

One of the typical device in the transmitter is the modulator.

momentarily 〔副〕瞬間的に

パイロットは、パワーレバーを押すことを一瞬ためらった。

The pilot momentarily hesitated to push the power levers.

momentary 〔形〕瞬間的な

風の変化は、パイロットに一瞬の困惑をもたらした。

Wind shift caused the pilot momentary confusion.

monitor 〔他〕監視する

RNAVルートの運航は、レーダーによる監視がなされていなければならない。

RNAV route operation must be monitored by radar.

Morse telegraphy 〔不〕モールス電信

航空無線通信士の免許は、モールス電信の運用は除外している。

The aeronautical radio operator's license excludes operation of Morse telegraphy.

multi 〔接頭辞〕多くの 種々の	MTSATは、多目的衛星である。 MTSAT is a multi purpose satellite.
must〔助〕〜ねばならない	機上の無線設備を操作する人は、航空無線通信士の免許を持っていなければならない。 A person who operates airborne radio equipment must have the aeronautical radio operator's license.

N

nationality 〔可·不〕国籍	航空機の国籍は、登録記号に示されている。 The nationality of an aircraft is shown by the registration mark.
nature 〔可·不〕自然	われわれは自然の汚染を最小に止めるべきである。 We should minimize pollution of nature.
natural 〔形〕自然の	無線周波数は、天然資源である。 Radio frequencies are natural resources.
navigation 〔不〕航法 航海	VOR/DMEを使った航法精度は優れている。 Navigation accuracy by means of VOR/DME is excellent.
near 〔形〕近い 〔前〕〜に近く	前線が空港に近いようである。 The frontal system may be near the airport.
nearby 〔形〕近くの 〔副〕近くで	パイロットは、どの局でも良いから近くの局と連絡をとる。 The pilot contacts any nearby station.
necessary 〔形〕必要な	急な風の変化に対する準備をすることは必要である。

N

It's necessary to prepare for a sudden wind shift.

---

need
〔可·不〕必要
強い向かい風のため余分な燃料が必要である。
There is a need of extra fuel for strong headwind.

〔他〕必要とする
国内運航の航空機には、HF通信装備は不要である。
Aircraft of domestic operation do not need HF communication equipment.

---

negate 〔他〕無効にする
前のクリアランスは、無効となった。
The previous clearance has been negated.

negative 〔形〕否定の
パイロットは、否定の応答をした。
The pilot responded with negative answer.

---

night 〔可·不〕夜
night effect 夜間効果
NDBの夜間効果により、ADFの精度は低下する。
Night effect of NDB causes deterioration in accuracy of ADF.

---

no longer もはや～でない
前線は通り過ぎた、もはや突風はない。
The frontal system has passed, gusty wind no longer exists.

---

noise 〔可·不〕騒音
noise abatement procedure 騒音軽減方式
空港における騒音を少なくするために、騒音軽減方式が推奨されている。
Noise abatement procedures are recommended to reduce noise at the airport.

---

non directional radio beacon (NDB) 無指向性無線標識
NDBは、航空路にも設備されている。
NDB is located on an airway as well.

---

non-revenue flight
非有償飛行
訓練飛行は、一つの非有償飛行である。
A training flight is one of the non-

N

revenue flights.

normal 〔形〕標準の 正常な
標準操作からの逸脱は、隠れた事故へと進む。
Deviation from the normal operation leads to a hidden accident.

normally 〔副〕正常に
エンジンは、正常に作動している。
The engine is working normally.

now and then 時々
空港は、雲の間から時々見え始めた。
The airport became visible now and then through the clouds.

number
〔可･不〕数 数字 番号
航空機の登録番号
aircraft registration number

numeral 〔可〕数字
アラビア数字　Arabic numerals

numerical
〔形〕数字で表した
気象状態は、数字のデータで表現される。
Weather condition is expressed as a numerical data.

O

object 〔可〕目標 目的
objective 〔可〕目標 目的
〔形〕目標の
標準化の目的は、安全運航の確保にある。
Object of standardization is to secure flight safety.

oblige
〔他〕義務を負わせる
無線従事者は、該当する規則に従って通信を行う義務がある。
Radio operators are obliged to communicate in compliance with the applicable regulations.

observe
〔他〕観察する 遵守する
パイロットは、管制官の指示を遵守しなければならない。
Pilots must observe controllers' instruction.

obstruction 〔不〕妨害
正規の状態で行われている通信に対する妨害は、有害な混信である。

Obstruction to the normally operated communication is harmful interference.

---

obtain 〔可〕得る

obtainable 〔形〕得られる 入手できる — 飛行場の気象状態は、ATISで入手できる。
Terminal weather is obtainable through ATIS.

---

obvious 〔形〕明らかな

obviously 〔副〕明らかに — 間違った周波数でコンタクト出来ないのは当然である。
It's obvious that contact is impossible on a wrong frequency.

---

occur 〔自〕起こる — 注意は、事故の発生を防止する。
Attentiveness will prevent an accident to occur.

occurrence 〔不〕発生 — ここでは雷雨は、稀にしか起こらない。
The occurrence of thunderstorms is rare here.

---

ocean 〔不〕大洋 — 国際線は、洋上を飛行する。
An international flight flies over the ocean.

oceanic 〔形〕大洋の — 海洋性気候は、穏やかである。
An oceanic climate is mild.

oceanic control area 洋上管制区域 — 洋上管制区域の交通は、ADSによって監視されている。
Traffic in the oceanic control area is monitored by ADS.

oceanic route surveillance radar (ORSR) 洋上航空路監視レーダー — ORSRは、二次レーダーである。
ORSR ia a secondary radar.

---

once 〔副〕一度 — 機は、一、二回バウンドして着陸した。
The aircraft bounded onece or twice at landing.

---

only 〔形〕唯一の	機長の唯一のコメントは、"注意すること"であった。 The captain's only comment was "be attentive".
〔副〕わずか	飛行時間は、わずか50分であった。 The flight time was only 50 minutes.
open 〔自·他〕開く	飛行中操縦室扉は、誰も開けてはならない。 No one can open the cockpit door in flight.
〔形〕開いた	扉を開けたままにしない。 Don't leave the door open.
operate	
〔自〕作動する	VORは、VHFで作動する。 VOR operates on VHF.
〔他〕操縦する	自動操縦装置は、機を精密に操縦する。 Auto-pilot precisely operates the aircraft.
operation 〔不〕運行 作動	パイロットは、機の各系統の作動を十分理解している。 Pilots fully understand the operation of aircraft systems.
operator 〔可〕操縦者	パイロットは、無線の通信士でもある。 Pilots are radio operators as well.
option 〔可·不〕選択 選択肢	待機するか代替空港に行くかの選択がある。 There are two options, to hold or to divert.
optional 〔形〕随意の	RNAVでは、パイロットは随意のルートを飛行できる。 RNAV allows pilots to fly optional route.

O

orbit
〔可〕軌道
〔他〕軌道に乗って回る

軌道に乗せられた衛星は、地球を回る。
A satellite put into orbit rounds the earth.

orbital 〔形〕軌道の
orbital data 軌道情報

航空機地球局は、衛星から軌道情報を受信して位置を求める。
An aircraft earth station calculates its position by receiving orbital data of satellites.

order
〔可〕命令
〔他〕命じる

命令に従う　obey order
機長は、パイロットに離陸開始を命じた。
The captain ordered the pilot to commence takeoff.

orderly 〔形〕整然とした

ATCは、整然とした交通の流れを維持する。
ATC keeps orderly flow of traffic.

ordinance 〔可〕条例

人々は、条例に使われている用語に必ずしも慣れてはいない。
People are not always familiar with the terms used in an ordinance.

organization
〔可〕組織体
〔不〕機構

その組織体は、運航に関っている。
The organization is engaged in flight operation.

organize 〔他〕組織する

会社は、通信網を組織化する。
The company organizes communication network.

origin 〔可·不〕発信元

通報の発信元が、不明である。
The origin of the message is unknown.

original 〔形〕初期の

初期の機上気象レーダーの操作は、難しかった。
The original airborne weather radar was difficult to operate.

originate 〔自･他〕始まる 始める	電気通信は、モールス電信から始まった。 Telecommunication originated with Morse telegraphy.
origination 〔不〕原点	
originator 〔可〕創始者	サンテグジュペリーは、航空の創始者と言える。 Antoine de Saint-Exupery is highly thought of him being an originator of aviation.
oscillate 〔自･他〕 振動する 振動させる	送信機内の発振器は、高周波の搬送波を発生する。 The oscillator in the transmitter generates high frequency carrier wave.
oscillator 〔可〕発振器	
other 〔形〕他の	他の周波数で連絡せよ。 Contact with other frequency.
otherwise 〔副〕さもなければ	私は、RVRが良ければ着陸するが、そうでなければ代替空港に行く。 I'll land if RVR is good, otherwise divert to the alternate airport.
outline 〔可〕概要 〔他〕概説する	パイロットは、進入計画の概要を簡潔に説明した。 The pilot gave a brief outline of his approach plan.
own 〔形〕自分自身の	自分自身の判断をしなさい。 Make your own decision.

P

park 〔他〕駐機する **parking spot** 駐機場	駐機場18番に駐機しなさい。 Park the aircraft at the parking spot 18.
part 〔可〕一部分 部品	長時間の巡航は、飛行の楽しい部分である。 Long cruise is an enjoyable part of a flight.
parts 〔可〕部品	整備員は､機の着陸前に部品を準備した。

P

The maintenance person arranged parts before arrival of the aircraft.

participate〔自〕参加する
そのクルーは、燃料節減協議に参加した。
The crew participated in a discussion of fuel economy.

particular〔形〕特有の
この機体は、右に傾く特有の傾向がある。
This aircraft has a particular tendency to bank to the right.

particularly〔副〕特に
速度のモニターは、特に重要である。
Speed monitoring is particularly important.

particularity
〔可･不〕特殊性
HF通信の特殊性は、消費電力が少ないことである。
Particularity of HF communication is low power consumption.

pass
〔自〕通過する
次の報告地点は、10時に通過するだろう。
We'll pass the next reporting point at 10:00.

〔他〕通り過ぎる
我々は、報告地点を10時に通過した。
We've passed the reporting point at 10:00.

passenger
〔可〕乗客
〔形〕旅客の
乗客の搭乗は終了した。
All passengers have been on board.
客室乗務員は、乗客用の救急箱をチェックした。
The cabin attendant checked the passenger first aid kit.

path〔可〕経路 コース
管制官は、雲を避けるように経路を修正した。
The controller corrected the path to avoid a cloud.

P

pattern 〔可〕型

管制官は、ホールディングパターンへと誘導した。

The controller guided the pilot to a holding pattern.

pause
〔可〕一時的休止 ポーズ
〔自〕一時的に休止する

区切りの無い通報は、受信者を混乱させる。

A message without a pause confuses the receiver.

penetrate 〔他〕貫通する

機上気象レーダーの電波は、積乱雲の氷結部分を貫通する。

The radar beam of the airborne weather radar penetrates the ice crystal zone of a cumulonimbus.

penetration 〔不〕貫通

貫通は、電波の持つ一つの特質である。

Penetration is one of the characteristics of a radio wave.

perfect 〔形〕完全な

澄み渡った青空の素晴らしいフライトであった。

The flight was perfect with clear blue sky.

perfectly 〔副〕完全に

滑走路は、完全に除雪されている。

The snow on the runway is perfectly removed.

periodic 〔形〕定期の

飛行場気象の定期的観測は、必要に応じ追加観測により補足される。

Periodic observation of terminal weather is supplemented by additional observation when necessary.

periodically
〔副〕定期的に

ATIS情報は、定期的にアップデイトされる。

ATIS information is periodically updated.

permit
〔自〕許す 可能にする

もし天気が良ければ、湾の方から滑走

P

	路へアプローチしよう。 Let's approach the runway from the bay side, if weather permits.
〔他〕許可する	操縦室では、喫煙は許可されていない。 Smoking is not permitted in the cockpit.
permission 〔不〕許可	操縦室に入ることは、許可されない。 No permission to enter the cockpit will be granted.
person 〔可〕人	彼らは、何人かの人が滑走路上を歩いているのを見た。 They saw some persons walking on the runway.
personnel 〔不〕職員	地上職員は、除雪で忙しい。 Ground personnel are busy with snow removal.
phase 〔可〕段階	着陸は、最も緊張する飛行段階である。 Landing is a flight phase the most tensional.
phrase 〔可〕熟語 慣用句	
phraseology 〔不〕用語 述語 専門語	法律用語には、厳密な意味がある。 Legal phraseology has the strict sense of the word.
piece 〔可〕破片 部品	パイロットが滑走路上に一個の異物を認めたので、離陸は中断された。 Takeoff was aborted because the pilots found a piece of foreign matter on the runway.
pilot 〔可〕操縦士 水先案内人	パイロットは、離陸中機の操縦に集中する。 Pilots concentrate themselves on aircraft maneuvering during takeoff.
pilot in command (**PIC**)	パイロットは、観察した悪天候を

機長
pilot report （**PIREP**）
機上気象報告

PIREPで報告するように要求されている。
Pilots are requested to report significant weather that they have observed by PIREP.

---

plan

〔可〕計画
全てのIFR飛行には、飛行計画をファイルしなければならない。
A flight plan must be filed for all IFR flights.

〔他〕計画する
運航管理者は、高速巡航を計画した。
The dispatcher planned a high speed cruise.

---

point

〔可〕点 小数点 地点
"point"とは、ある場所のことである。
The point is a location of certain place.

〔自〕示す
IASの指針は、320ノットを示している。
The IAS indicator needle is pointing to 320kt.

---

policy
〔可･不〕方策 指針 手段
唯一の運航の方策は安全である。
The only flight operation policy is safety.

---

portion 〔可〕一部分
このフライトの最初の部分では、通常向かい風が強い。
The headwind component is usually strong for the first portion of this flight.

---

position

〔可〕位置 場所
パイロットは、離陸位置でクリアランスを待った。
The pilot waited for the clearance at the takeoff position.

〔他〕置く
PAPIは、接地帯の側に置かれている。
PAPI is positioned at the side of

P

	the touchdown zone.
position report 位置通報	パイロットは、国際線においては位置通報に気象情報を付け加える。 Pilots add weather information to the position report in case of an international flight.
possibility 〔不〕可能性	天候回復の可能性は、期待できる。 Possibility of weather recovery is hopeful.
possible 〔形〕可能な	満席の乗客で12時間以上飛行することは、可能である。 It is possible to fly longer than 12 hours with full passenger.
power 〔不〕出力 動力 電力 電気	無線機器による消費電力は、少ない。 Power consumption by radio equipment is less.
power supply 電力供給	航空機の通常の電力供給は、エンジン駆動の発電機で賄われている。 Normal power supply of an aircraft is covered by engine driven generators.
practical 〔形〕実践的な	実際の経験により技量が上達する。 A practical experience improves skill.
practice 〔不〕実践 練習	新人パイロットは、頻繁に離着陸を実践すべきである。 New pilots should have frequent takeoff and landing practice.
pre-departure 出発前 **pre** 〔接頭辞〕時間的、位置的に（前に）の意	飛行計画と天候の全体について、出発前のブリーフィングで説明された。 Overall flight plan and weather condition were explained in the pre-departure briefing.
precede 〔他〕先んずる	先行機は、最終進入5マイルの点にいる。 The preceding traffic is at 5 miles on final.

P

precise 〔形〕精密な 的確な	DMEは、精密な距離測定を行う。 DME measures precise distance.
precisely 〔副〕精密に	精密に測定された距離 precisely measured distance
precision 〔不〕精密 〔形〕高精度の	高精度進入方式 precision approach procedure
precision approach radar (PAR) 精密進入レーダー	PARは、機をGCAへと誘導する。 PAR guides an aircraft to GCA.
preface 〔可〕序文	序文は、本の最初の頁にある。 The preface is on the first page of a book.
prefer 〔他〕～の方を好む	航空会社のパイロットは、GCAよりもILSアプローチの方を好む。 Airline pilots prefer ILS approach to GCA.
preferable 〔形〕 （～より）より望ましい	他の空港へ行くよりも待機の方が好ましい。 Holding is preferable to diversion.
preferential 〔形〕優先の	エンジン故障の場合には、ATCの優先措置がとられる。 ATC preferential handling will be given to an aircraft with an engine failure.
prepare 〔他〕用意する 準備する	パイロットは、急な風の変化に対して準備していなかった。 The pilot was not prepared for a sudden wind shift.
preparation 〔可·不〕用意 準備	地上員は、プッシュバックの準備を整えた。 The ground crew has completed preparation for push back.
prescribe 〔他〕規定する	緊急降下手順は、AOMに規定されている。 The emergency descent procedure is prescribed in the aircraft operating manual (AOM).

P

present 〔形〕現在の

現在の機上の受信機には、SELCALが装備されている。

The present airborne receivers are equipped with SELCAL system.

prevent 〔他〕予防する

無線操作者は、混信を予防するために不必要な送信をすべきでない。

A radio operator should refrain from unnecessary transmission to prevent interference.

previous 〔形〕以前の

風のデータは、前の情報より大いに異なる。

The wind data is quite different from the previous information.

previously 〔副〕以前に(は)

彼は、以前プロペラ機で飛んでいた。

He had previously flown propeller aircraft.

primary 〔形〕初期の 主要な 一次の

レーダーは、一次レーダーと二次レーダーに区分されている。

The radar system is classified into the primary radar and the secondary radar.

primary radar 一次レーダー

機上の気象レーダーは、一次レーダーである。

The airborne weather radar is the primary radar.

principle 〔可〕原理 原則

ATC通信の原則は、世界規模の標準化である。

The principle of ATC communication is world wide standardization.

prior 〔形〕前の

prior to 〜より前に

機長は、離陸に先立って確実な速度制御を強調した。

The captain emphasized precise speed control prior to takeoff.

priority 〔不〕優先順位

遭難通信は、絶対優先である。

The distress traffic has absolute priority.

priority handling 優先措置

緊急状態に対する優先措置

priority handling for an emergency

procedure 〔可·不〕手順 方式	無線従事者は、ICAOの手順に従うべきである。 A radio operator should follow the ICAO procedures.
proceed 〔自〕進む 続行する	機は、代替飛行場へと行った。 The aircraft proceeded to the alternate airport.
process 〔可·不〕過程 経過 〔他〕処理する	FMSのデータ処理能力によりパイロットは、複雑な頭脳労働から解放された。 Data processing capability of FMS has relieved pilots from complicated brainwork.
produce 〔不〕産物 〔自·他〕製造する	AM方式は、サイドバンドを発生する。 The AM method produces side bands.
progress 〔不〕進歩 経過 〔自〕進歩する	飛行経過は、正常である。 Flight progress is normal.
promote 〔他〕 促進する 助長する 増進する	注意深い計画は、安定したアプローチを増進する。 Careful plan will promote stabilized approach.
prompt 〔形〕迅速な	パイロットには、迅速な行動が要求される。 A prompt action is required to be taken by a pilot.
〔他〕鼓舞する 刺激する	近づく霧の情報にパイロットは、急いでアプローチに入った。 The pilot was prompted by the news of nearing fog to proceed to approach.
promptly 〔副〕迅速に	パイロットは、速やかに経路を修正した。

P

The pilot promptly corrected the path.

pronounce 〔他〕発音する

ICAOは、用語や数字の発音は既に発行してあるとおりに行うよう推奨している。

ICAO recommends pronunciation of terms and numbers as it is published.

pronunciation
〔不〕発音

言語習慣は、発音の相違をもたらす。

Language habit makes difference in pronunciation.

propagate 〔他〕伝播する 普及させる 繁殖させる

電波は、媒介物無しで遠くへ伝播する。

Radio waves propagate far distance without a vehicle.

propagation
〔不〕伝播 普及 繁殖

電波は、伝播の過程でいろいろな特徴を示す。

Radio waves give various characteristics during propagation.

proportion
〔不〕比率 割合 調和
in proportion to
~に比例して

残存燃料は、残りの距離に比例して減少している。

Remaining fuel has been decreasing in proportion to the distance to go.

propose
〔他〕企てる 提案する

我々は、飛行計画で高度39,000フィートを提案した。

We proposed FL390 in the flight plan.

provide
〔自･他〕提供する 規定する

航空会社は、長距離飛行の場合食事を提供する。

Airlines provide passengers with food for a long range flight.

provision
〔不〕提供 規定

禁煙に関する条項は、改定されている。

The provisions related to “no smoking” have been revised.

proximate 〔形〕最も近い
機は、滑走路に最も近い停止線で止まらなければならない。
An aircraft must stop at the holding line proximate to the runway.

proximately 〔副〕最も近く

pseudo 〔形〕擬似の
pseudo station 擬似局
GPSの航法精度は、擬似局の追加によって改善された。
GPS navigation accuracy has been improved by adding a pseudo station to the system.

publicize 〔他〕公表する
進入手順の詳細は、チャートにより公表されている。
Detail of approach procedure is publicized by the chart.

publication 〔可·不〕出版 出版物
AIPは、政府刊行物である。
AIP is a government publication.

publish 〔自·他〕出版する 公表する
飛行解析報告は、隔月に発行される。
The flight analysis report is published every other month.

pull
〔可〕一引き
操縦桿の一引き
a pull of the control column
〔不〕引力
月の引力 the pull of the moon
〔自·他〕引く
パイロットは、減速のためにパワーレバーを引いた。
The pilot pulled the power levers to reduce speed.

pulse
〔可〕波動 パルス
〔自〕鼓動する
パルスとパルスの間は、受信期間である。
The space between pulses is a receiving period.
pulse modulation（PM） パルス変調

purpose
〔可〕目的 意図
〔他〕意図する
シミュレーターは、非常時操作手順の訓練目的にとって有効である。
A simulator is effective for the purpose of emergency procedure training.

push-back
〔不〕〔他〕プッシュバック

プッシュバックは、地上員によって行われる。
Push-back is conducted by a ground crew.

Q

qualify 〔自・他〕
資格を取る 資格を与える

私は、来月資格を取る。
I will qualify next month.
あなたは航空会社のパイロットの資格を与えられる。
You will be qualified as an airline pilot.

qualification 〔可〕資格

無線従事者は、該当する資格を持っていなければならない。
A radio operator must have an appropriate qualification.

quality 〔不〕品質

衛星経由のVHF長距離通信の品質は、HFシステムよりはるかに良い。
VHF long range communication quality through a satellite is much better than the HF system.

question
〔可〕質問 質疑

パイロットは、機長に難しい質問をした。
The pilot put a difficult question to the captain.

〔不〕疑問 疑い

もうじき霧は疑いなく晴れる。
There's no question that the fog will clear soon.

〔他〕質問する

指導者は、彼に天文航法について質問した。
The instructor questioned him on celestial navigation.

R

radar
〔可〕電波探知機
レーダー装置

国内交通は、レーダーで監視されている。
Domestic traffic is monitored by radar.

〔形〕レーダーの（radio detecting and ranging）

RADARの語は、無線による探知と測位の略である。
Radar stands for “radio detecting and ranging”.

radar approach
レーダーアプローチ

交通の流れは、レーダーによる進入で促進される。
Traffic flow is expedited by radar approach.

radar contact
レーダーコンタクト

管制官は、特定の機をレーダーで捕捉し識別した。
The controller identified a specific aircraft by radar contact.

radar coverage
レーダー覆域

機上気象レーダーの覆域は、選択できる。
The radar coverage of an airborne weather radar is selectable.

radar echo
レーダーエコー

レーダーエコーを読み取るには、電波の知識を必要とする。
Reading a radar echo requires knowledge of radio waves.

radar environment
レーダーの環境

レーダー環境は厳しい。そのため雨が目標を遮蔽することもある。
Radar environment is severe. Therefore the rain may block the target.

radar equipment
レーダー装置

異なる種類のレーダー装置が、航空交通の管制に使われている。
Different type radar equipment is applied to control air traffic.

radar service
レーダーサービス

レーダーサービスは、VMCの航空機にも提供される。
Radar service is available even to VMC aircraft.

---

radio

〔可〕無線装置
無線受信機　radio receiver

〔不〕無線電信 放送
船は、無線電信で通信する。
The ship communicates by radio.

〔形〕無線の
無線連絡　radio contact

〔他〕（通信を）無線で送る
船は、気象情報を無線で送る。

The ship sends weather information by radio.

radio beacon 無線標識
無線標識は、識別信号を放射する。
A radio beacon emits an identification signal.

radio frequency 無線周波数
緊急無線周波数は、121.5MHzである。
The emergency radio frequency is 121.5MHz.

radio navigation 無線航法
無線航法により正確にコース上を飛行することが可能になった。
Radio navigation has made it possible to precisely fly on course.

radio navigation facility 無線航法施設
無線航法施設は、機の位置測定に使われる。
Radio navigation facilities are used to determine aircraft position.

radio operator 無線従事者
無線従事者は、他国上空での無線機器の操作が許される。
A radio operator may operate radio equipment over the foreign country.

radio law 電波法
無線通信は、電波法の条項に従って実施されるべきである。
Radio communication must be conducted in compliance with the articles of the Radio Law.

radio telegraphy 無線電信
無線電信は、ATCにおいては会話式の無線通信によって取って代わられた。
Radio telegraphy was superseded with conversational radio communication in ATC.

radio wave 電波
通信に使う電波の周波数は、3kHzから300GHzまでである。
Radio wave frequencies that are used for radio communication range from 3kHz to 300GHz.

R

radius 〔可〕半径	飛行場標点から9キロメートルの半径で3,000フィートまでの空域が、管制圏である。 The area of 9km radius of the airport reference point and up to 3,000ft is the control zone.
range 〔不〕範囲 距離	ASRの有効範囲は、60マイルである。 The effective range of ASR is 60nm.
〔動〕及ぶ	機上気象レーダーの最大覆域は、約300マイルに及ぶ。 The maximum coverage of the airborne weather radar ranges approximately 300 nautical miles.
rapid 〔形〕素早い **rapidly** 〔副〕速やかに 急速に	素早い反応 rapid response 航空交通密度は、急速に増加した。 Air traffic density has rapidly increased.
rate 〔可〕率 進度	ピッチ上げ下げの率は、対気速度に微妙に影響する。 Pitch rate delicately acts on the air speed.
rational 〔形〕合理的な **rationally** 〔副〕合理的に	合理的な取扱い rational handling 合理的に取扱う rationally treat
reach 〔他〕届く 到達する	VHFの電波は、障害物の後ろ側にも届く。 VHF radio waves reach even behind an obstacle.
read back 〔不〕復唱 〔自·他〕復唱する	パイロットは、復唱により通報を確認する。 The pilot confirms the message by read back.
readability 〔不〕読取りやすさ	パイロットは、ラジオチェックにより読取り具合を確かめる。

	The pilot confirms readability by a radio check.
readable 〔形〕読取り可能な	そちらの通信は、読取り可能である。 Your message is readable.
ready 〔形〕準備が整って	離陸の準備は整っている。 We are ready for takeoff.
realize 〔他〕理解する 実現する	パイロットは、天気が限界に近いことを理解した。 The pilot has realized the weather is marginal.
reason 〔可·不〕理由	機長は、目的地変更の理由を乗客に説明した。 The captain explained the reason for diversion to passengers.
receive 〔自·他〕受取る 受信する	パイロットは、出発遅延の情報を受け取った。 The pilot received a message of departure delay.
receipt 〔可〕領収	受信証 an acknowledgement of receipt
receiver 〔可〕受信機	機上の受信機には、SELCALが付いている。 The airborne receiver is equipped with SELCAL.
receiving station 受信局	パイロットは、受信局での受信を容易にするために、通報をはっきり読む。 The pilot reads the message clearly to facilitate reception by the receiving station.
recognize 〔他〕認める **recognition** 〔不〕認識	飛行データ分析者は、やや遅れ目のフレアーがハードランディングの原因であることを認めた。 The flight data analyst recognized slight delay in flare caused hard landing.

R

recover
〔自·他〕立直る 回復する
天気が回復するまで待機することは出来ない。
Holding until the weather recovers is impossible.

recovery 〔不〕回復
天気の回復には時間がかかる。
Weather recovery will take time.

rectify 〔他〕調整する
rectifier 〔可〕整流器
交流は、整流器で直流に整流される。
AC is rectified to DC by a rectifier.

reduce〔自・他〕縮小する
200ノットに減速せよ。
Reduce speed to 200kt.

reduction 〔可·不〕縮小
会社は、フリートの縮小を発表した。
The company announced reduction in fleet size.

redundant 〔形〕冗長な
冗長な表現は、理解し難い。
Redundant expression is difficult to understand.

redundancy 〔不〕冗長
システムの冗長性は、安全な作動を向上させる。
System redundancy enhances safe operation.

refer 〔自〕参照する
reference
〔可〕参照文
〔不〕参照
〔形〕参照の
この本は、主として公の刊行物を参照している。
This book mainly refers to the official publications.
参考資料は、報告書の末尾にある。
Reference material is attached to the end of the report.

reference phase
基準位相
VORの方位は、基準位相と可変位相の位相差から求められる。
VOR determines the bearing from the phase difference of the reference phase signal and the variable phase signal.

reflect〔自·他〕反射する
F層は、HFの電波を反射する。
F layer reflects HF radio waves.

reflection
スポラジックE層は、VHFの反射の原

R

〔可〕映像 〔不〕反射

因となる。

Sporadic E layer causes reflection of VHF radio waves.

---

refrain

〔自〕控える 我慢する 断つ

無線従事者は、不必要な送信をしてはならない。

A radio operator should refrain from unnecessary transmission.

R

---

regard

〔不〕顧慮 関心

〔他〕～とみなす

パイロットは、さらに多く安全に気を配る。

Pilots pay more regard to safety.

事態は、重大視されている。

They regard the situation as serious.

regardless

〔形〕不注意な 不注意で

〔副〕～にかまわず

自動着陸では、視程が悪くても安全な着陸が出来る。

Auto-land enables safe landing regardless of poor visibility.

---

register

〔可〕登録 登録簿

〔他〕登録する

その航空機は、日本の航空局に登録されている。

The aircraft is registered with JCAB.

registration

〔可〕登録証明書

〔不〕登録 書留

その航空機の登録記号は、JA6001である。

The registration mark of the aircraft is JA6001.

---

regular

〔形〕規則的な 定期の 公認の

パイロットは、所定の目視経路に従って滑走路へと進入した。

The pilot approached the runway by following a regular visual pattern.

regularity

〔不〕秩序 調和

ATCは、交通流の秩序を整える。

ATC sets regularity of traffic flow.

regulate

〔他〕規制する 調整する

管制官は、進入する航空機の着陸の順序を調整する。

The controller regulates landing sequence of approaching aircraft.

regulation 〔可〕規則 規定 法規	機の取扱要領は、該当する規定に詳しく述べてある。 The aircraft operating procedures are described in detail in the appropriate regulation.
reject 〔他〕拒絶する	パイロットは、エンジン故障のため離陸を止めた。 The pilot rejected takeoff because of an engine failure.
rejection 〔可·不〕否決 拒絶	離陸中断は、直ちにタワーに報告された。 Takeoff rejection was immediately reported to the tower.
relate 〔自〕関連がある 〔他〕関係させる	この問題に天候は関係していない。 The weather does not relate to this trouble.
relative 〔形〕相関的な	信号対雑音の比率は、受信機の品質と相関的である。 Signal-to-noise ratio is relative to the receiver quality.
relation 〔可〕関連 間柄 〔不〕関係	天候と飛行の快適さの関係 The relation between weather and comfortableness of flight
relay 〔可〕中継 〔他〕中継する	伝達席は、パイロットに管制承認を中継する。 The delivery section relays pilots the clearance.
reliability 〔不〕信頼度	エンジンの信頼度は高い。 Reliability of an engine is high.
reliable 〔形〕信頼できる	FM受信機は、音の清澄さで信頼できる。 FM receiver is reliable with sound clarity.
remain 〔自〕残存する	残存燃料は、十分である。 Remaining fuel is sufficient.

R

render 〔他〕与える	その給油基地は、必要なサービスを全て提供する。 The refueling station renders all necessary service.
repeat 〔可〕繰返し 〔他〕繰返す	パイロットは、復唱の中でクリアランスを繰返した。 The pilot repeated the flight clearance in his read-back.
repetition 〔可·不〕繰返し 反復	彼の反復は、正しかった。 His repetition was correct.
replace 〔他〕交換する 取替える	整備員は、作動装置を交換した。 The maintenance person replaced the actuator.
reply 〔可〕回答 〔自〕応答する 〔他〕返答する	受信局は、即座に応答した。 The receiving station promptly replied.
reply procedure 応答手順	操作者は、所定の応答手順に従うべきである。 The operator should follow the pre-set reply procedures.
report 〔可〕報告 〔自·他〕報告する	パイロットは、通信に困難があった場合報告を提出する。 A pilot will submit a report in case communication difficulties were experienced.
request 〔可·不〕要請 〔他〕要請する	パイロットは、雲を避けるためにコースからの逸脱を要請した。 The pilot requested deviation from the course to avoid a cloud.
require 〔他〕必要とする	37,000フィートから20,000フィートまで降下するには、約10分必要である。 It requires around 10 minutes for descent from 37,000 feet to 20,000 feet.

requirement 〔可〕必要条件 必要品	離着陸に必要な気象要件は、最低気象条件といわれる。 Weather requirement for takeoff and landing is called the "weather minima".
rescue	
〔不〕救出 救助 〔他〕救出する	火災の航空機から全ての乗客が救出された。 All passengers were rescued from the burning aircraft.
〔形〕救難の レスキューの	レスキュー隊　rescue party 救難作業　rescue work
rescue coordination center (RCC) 救難調整本部	RCCは、直ちに設立された。 RCC was immediately established.
resolution advisory (RA) 回避情報	RAは、予測衝突時の25秒前に作動する。 RA is activated 25 seconds prior to predicted collision point.
resonate 〔自〕共鳴する	アンテナは、素子の長さが周波数に適合しているときに共鳴する。 An antenna resonates with a frequency when the element length is appropriate to it.
resource 〔可〕資源 力量 **human resources** 人的資源 **natural resources** 天然資源	パイロットは、操縦室における人的資源である。 Pilots are human resources in the cockpit.
respective 〔形〕各自の それぞれの **respectively** 〔副〕それぞれ	機長と操縦士はそれぞれの席に着いた。 The captain and the pilot took their respective seat.
respond 〔自･他〕応答する 返答する **response** 〔可･不〕返答 応答	SELCALが不作動であったので、パイロットは返答できなかった。 SELCAL was inoperative, and the pilot failed to respond.

responsibility
〔可〕（具体的な） 責任
〔不〕責任 義理

管制官の責任は、非常に重要である。
A controller's responsibility is a matter of critical importance.

rest
〔不〕休憩 静止 停止

１時間の休憩　an hour for rest
機は、止まった。
The aircraft has come to rest.

〔自･他〕休む ～にかかる

ベッドに行って休みなさい。
Go to bed and rest.
着陸の可能性は、霧の晴れる時間にかかっている。
The possibility of landing rests with the time of the fog to clear.

restrict
〔他〕制限する 限定する

進入区域における10,000フィート以下の速度は、250ノットに制限されている。
The speed below 10,000 feet in the approach area is restricted to 250kt.

restriction
〔可･不〕制限 限定

速度制限は、管制官の許可により解除され得る。
The airspeed restriction may be cancelled by the controller's permission.

resume
〔自･他〕再開する 回復する

機は、通常の航法に戻った。
The aircraft resumed normal navigation.

revenue flight
〔可･不〕有償飛行

訓練飛行は、有償飛行ではない。
Training flight is not a revenue flight.

review 〔可･不〕再調査
〔自･他〕再調査する よく調べる

ACCは、飛行計画をよく調べて、管制承認を発行する。
ACC reviews a flight plan and issues a flight clearance.

roll 〔自〕転がって進む

takeoff roll 離陸滑走 **landing roll** 着陸滑走	パイロットは、離陸滑走中機の操縦に専念する。 Pilots concentrate themselves in maneuvering the aircraft during takeoff roll.
route 〔可〕航路 航空路 幹線道路	長距離飛行のルートは、大圏コースを基準にしている。 The route of a long range flight is based on a great circle course.
rule 〔可〕規則 通例 〔自･他〕支配する	操縦室への立入り禁止は、規則によって決められている。 It's a rule that an entry to the cockpit is prohibited.
runway 〔可〕滑走路	滑走路は、遠くから見えていた。 The runway was visible from a distance.
runway visual range (RVR) 滑走路視距離	RVRは、滑走路上3点で測定される。 RVR is measured at three points on the runway.

S

safe 〔形〕安全な	航空会社の全員が、安全運航を目指す。 The entire staff of an airline aims safe flight.
safety 〔不〕安全 〔形〕安全のための	飛行の安全　flight safety 誤操作を防ぐ安全装置が適用されている。 A safety device to protect inadvertent operation is applied.
same 〔形〕同一の 同様の	NDB局と放送局の電波の周波数帯は同じである。 The frequency band of an NDB station and a broadcasting station is same.
satellite 〔可〕衛星 **weather satellite** 気象衛星	MTSATは、多目的衛星である。 MTSAT is a multi purpose satellite.

satisfy 〔他〕満足させる	飛行計画は、機長を満足させるものであった。 The flight plan satisfied the captain.
satisfactory 〔形〕申し分の無い 満足な	パイロットは、飛行を満足な状態で実施した。 The pilot managed the flight in a satisfactory manner.
satisfactorily 〔副〕思うとおりに 満足に 十分に	地上業務は、ちゃんと提供された。 Ground handling was satisfactorily provided.
scale 〔可〕スケール	航空の時間尺度は、世界協定時である。 The aviation time scale is in UTC.
scatter 〔不〕撒き散らされた状態 〔自･他〕散らす	風が霧を散らした。 The wind scattered the fog.
〔不〕少量	少量のアイスパッドが、滑走路上にあった。 A scatter of ice pad remained on the runway.
schedule 〔可〕予定（予定表）	ATISの予定変更が、NOTAMで発表されている。 An ATIS schedule change is announced by the NOTAM.
〔他〕予定する	耐久飛行は、明日に予定されている。 The endurance flight is scheduled for tomorrow.
scope 〔可･不〕範囲	MTSATの機能の範囲は、航空ミッションと気象ミッションに及ぶ。 The functional scope of MTSAT extends to the flight mission and the weather mission.
screen 〔可〕スクリーン	ASRとSSRの両方の情報がレーダースクリーン上に表示される。 Both ASR and SSR information is displayed on the radar screen.

S

search
〔可〕捜索
〔自〕調査する
〔他〕捜す
search and rescue
捜索救難

行方不明機の捜索が行われている。
The search for the lost plane is in progress.
日本は、福岡FIR内の捜索救難活動の責任を持っている。
Japan is responsible for search and rescue service within Fukuoka FIR.

second
〔可〕秒　〔形〕二番目の
secondary〔形〕第二位の

現在B747は、世界で二番目に大きい旅客機である。
Now, B747 is the second largest airliner in the world.

secondary radar
二次レーダー
secondary surveillance radar (**SSR**)
二次監視レーダー

SSRは、UHFで作動する二次レーダーである。
SSR is a secondary radar that operates on UHF.

section〔可〕部分

タワーは、三つの部分から成る。
The tower consists of three sections.

secure
〔他〕(～の安全を)確保する

着陸は確実だとパイロットは言った。
The pilot said the landing was secured.

security〔不〕安全 保障

油断大敵
Security is the greatest enemy.

segment〔可〕区分

通信衛星は、CNS/ATMの宇宙区分である。
The communication satellites are the space segment of CNS/ATM.

select〔他〕選ぶ

指定された（割当）周波数を選びなさい。
Select the assigned frequency.

selection〔可·不〕選択

パイロットは、自動着陸モードの選択を完了した。
The pilot completed the selection of auto-land mode.

S

S

selective calling system (SELCAL) 選択呼出装置	SELCALは、パイロットの作業負担を軽減する。 SELCAL reduces pilots' workload.
send〔他〕送る 発信する	管制官は、通報を一括呼出しで発信した。 The controller has sent the message on "all call".
sentence〔可〕文 文章	最近の技術報告には、平叙文がより多く使われる。 A declarative sentence is more popular in recent technical reports.
separate 〔自〕離れる 分離する	客室は、３クラスに分かれている。 The cabin separates into three classes.
〔他〕分離する 切り離す **separately**〔副〕別々に **separation**〔可·不〕分離	音声通信システムは、VHFとHFの二つのシステムに分かれている。 The voice communication system is separated into two systems, the VHF and the HF system.
serious〔形〕重大な **seriously**〔副〕重大に	その天候は、もし長引けば、パイロットにとって重大な困難となるかもしれない。 The weather if it lasts long might put the pilots in a serious difficulty.
service 〔不〕サービス 業務 運行 事業	航空放送業務は、航空通信業務の中の情報提供業務である。 The aeronautical broadcasting is an information service of an aeronautical communication system.
severe〔形〕厳しい **severely**〔副〕厳しく	冬の気候は、離着陸にとって厳しい。 Winter weather is severe on takeoff and landing.
shall〔助〕	（法律、規則の条文で）～すべきである

shallow 〔形〕浅い

パイロットは、浅い降下経路を選んだ。
The pilot selected a shallow descent path.

shift
〔可〕変化
〔自〕(風の) 向きが変わる

予測しない風の変化が起きた。
Unexpected windshift occurred.

shockwave 〔可〕衝撃波

機の速度が音速になると、衝撃波が発生する。
A shockwave is generated when the aircraft speed reaches the speed of sound.

short 〔形〕近い 短い
〔副〕（～の） 手前で

機は、滑走路手前の停止線で止まらなければならない。
Aircraft must stop at the stop line short of the runway.

short range 短距離

短距離機　a short range aircraft

show 〔他〕示す

機は、接地時に右への流れを示した。
The aircraft showed a drift to the right on touchdown.

sick 〔形〕病気の

パイロットは、病気の乗客のため予定外の着陸をしなければならなかった。
Pilots had to make an unscheduled landing due to a sick passenger.

sight 〔不〕視力 視覚 視界

パイロットにとって、滑走路は遠くから視界内にあった。
The pilot had the runway in sight from a distance.

sign
〔可〕合図 符号 前兆

黒雲は、雨の前兆である。
A dark cloud is a sign of rain.

〔他〕署名する

ディスパッチャーと機長が、飛行計画書に署名した。
The dispatcher and the captain signed the flight plan.

S

signal 〔可〕信号 〔他〕合図する	遭難信号のMAYDAYが緊急周波数で聞こえる。 MAYDAY, the distress signal, is heard on the emergency frequency.
signal wave 信号波	信号波が、搬送波を変調する。 The signal wave modulates the carrier wave.
signify 〔自･他〕意味する 表明する	滑り易い滑走路への着陸には、いろいろなことが関る。 Landing on the slippery runway signifies a lot.
significant 〔形〕意味深い 重要な 大きな影響を与える	重大な影響を与える気象情報 significant weather information (SIGMET)
silence 〔不〕沈黙	MAYDAYの語は、他の局の沈黙を命じる。 The word MAYDAY commands silence of other station.
similar 〔形〕類似した	類似した呼出符号の使用は、危険である。 Use of similar call signs are dangerous.
similarity 〔可〕類似点 〔不〕類似	STOL機とVTOL機には類似点がある。 There are some similarities between the STOL and VTOL aircraft.
simultaneous 〔形〕同時の 同時に起る **simultaneously** 〔副〕同時に	無線連絡とシステム操作とが同時に起ると、パイロットは忙しくなる。 Simultaneous occurrence of radio contact and system operation makes pilots quite busy.
simple 〔形〕簡単な	簡単な操作　simple operation
since 〔接〕〜してから 〜だから	旅客機の速度が、M.82になってから40年以上経っている。 It's more than 40 years since the airliners' speed had reached M.82.

sink
〔自〕沈む
〔他〕沈める
sink rate 沈下率

パイロットは、接地前に機の沈下を止める。
A pilot stops sinking of the aircraft before touch down.

situation
〔可〕立場 情勢 環境

航空輸送にとって、燃料の情勢は厳しい。
The fuel situation is severe for air transportation.

slant
〔可·不〕傾斜 斜面 傾向
〔形〕傾斜した
slant distance 斜距離

DMEで測定された距離は、斜距離である。
The distance measured by DME is a slant distance.

slight 〔形〕少しばかりの
slightly 〔副〕わずかに

霧ともやには若干の違いがある。
There's a slight difference between fog and mist.

smooth
〔形〕順調な 平坦な
smoothly 〔副〕難なく

パイロットは、雲の中を順調に降下した。
The pilot smoothly descended in clouds.

some 〔形〕多少の

落雷で胴体表面に多少の傷が出来た。
Some dents are found on the fuselage skin caused by a lightening strike.

somewhat 〔副〕多少

雲の中では多少は揺れる。
It's somewhat bumpy in a cloud.

sound
〔不〕音
〔自〕音を出す
〔他〕鳴らす

旅客機が音速以上で飛行することは、遠い将来であろう。
It'll be a distant future that an airliner flies faster than the speed of sound.

space
〔可·不〕空間 宇宙 空所
〔形〕宇宙の

いろいろな目的の多くの衛星が、宇宙で作動している。
Many satellites of various purposes are operating in space.

S

speak 〔自･他〕話す	ATC通信においては、パイロットも管制官も英語を話す。 Pilots and controllers speak English in ATC communication.
specify 〔他〕明示する	飛行計画に記載する項目は、特定されている。 Items to be listed in a flight plan are specified.
specific 〔可〕細目〔形〕特定の	各トランスポンダーコードには、特定の意味がある。 Each transponder code has a specific meaning.
spectrum 〔可〕スペクトル （変動の） 範囲	乗客用娯楽システムは、広範囲の趣味の要望を満たすことが出来るであろう。 The passenger entertainment system (PES) will meet the demand of a wide spectrum of interests.
speech 〔不〕話し方	ATC通信の話し方の速度は、ゆっくりであるべきである。 The speed of speech in ATC communication should be slow.
speed 〔可･不〕速度 **speed brake** スピードブレーキ **speed of sound** 音速 **indicated air speed**（**IAS**）指示対気速度 **true air speed** （**TAS**） 真対気速度 **ground speed** （**GS**） 対地速度	航空機の速度は、指示対気速度、真対気速度及び対地速度で表示される。 Aircraft speed is expressed in IAS, TAS and GS. 飛行には、IASを使う。 IAS is used for flight. 飛行計画には、TASを使う。 TAS is used for a flight plan. 乗客への情報には、GSを使う。 GS is used for passenger service.
spell 〔自･他〕綴る	単語を正しく綴りなさい。 Spell a word correctly.

spend
〔他〕費やす 過ごす 消耗する

パイロットは、待機に約30分費やすと考えた。
The pilot estimated to spend around 30 minutes in holding.

spot 〔可〕場所 地点

駐機場は、18番である。
The parking spot is No.18.

squawk
〔自〕やかましく不平を言う
（右の**2**例は、パイロットの慣習的用法）

トランスポンダーコード5678を作動させる。
Squawk transponder code 5678.
発電機の故障を知らせる。
Squawk an inoperative generator.

stabilize
〔自〕安定する
〔他〕安定させる

安定したアプローチは、安全な着陸となる。
Stabilized approach will contribute to a safe landing.

stabilization〔不〕安定

自動操縦と推力自動制御で機は安定している。
Aircraft stabilization is secured by the automatic flight control and the thrust control.

standard
〔可〕標準 基準
〔形〕標準の

マラソンの標準距離は、42.195kmである。
The standard distance of Marathon is 42.195km.

standard procedure
標準方式

一旦静止後離陸を開始する方法は、湿潤滑走路上での標準方式である。
Standing takeoff is the standard procedure on a wet runway.

standard operating procedure (SOP)
標準操作手順

航空会社のパイロット訓練の目標は、標準操作手順である。
The objectives of an airline pilot training are SOP.

standard instrument departure (SID)
標準計器出発方式

出発経路のSID部分は、通常レーダー誘導で飛行される。
SID portion of departure course is usually flown by radar vectoring.

standard terminal arrival route (STAR)
標準到着経路

STAR空域の交通流は、レーダー誘導により効率よくコントロールされる。
Radar vectoring efficiently controls traffic flow in STAR area.

standardization
〔可·不〕標準化 規格化 統一

航空機の各系統の機能及びそれらの統合に関する技術的進歩により、パイロットの各系統の操作手順の標準化が容易に行われた。
Technical advancement in aircraft system function and their integration facilitated standardization of pilot's system operating procedures.

S

---

standby
〔可〕交代要員
〔形〕予備の

（ICAO標準用語では、次の意味で使用する。）
当方から呼び出すまで待っていてください。
Wait and I will call you.

---

state
〔可〕国 州
〔他〕述べる

加盟国 member state
規則では、喫煙は禁止と述べている。
The regulation states that smoking is prohibited.

---

status 〔不〕地位 事情

航空輸送は、以前高く評価されていた。
The status of air transportation had been highly appreciated.

---

steep 〔形〕急勾配の

急な旋回や上昇は、乗客に不快感をもたらす。
A steep turn or clime contributes un-comfortableness to passengers.

---

stipulate 〔他〕規定する

"安定したアプローチ"の実施については運航規程に規定されている。
Operation of "stabilized approach" is stipulated in the operations manual.

---

stop
〔可〕停止 〔自〕止まる
〔他〕止める
stop line 停止線

機は、停止線で止まった。
The aircraft has come to a stop at the stop line.
パイロットは、緊急ブレーキで機を止めた。
The pilot stopped the aircraft using the emergency brakes.

subject
〔可〕科目 課題
〔形〕～を条件として

進入空域において公表速度を超過するには、管制官の許可を条件とする。
Exceeding the published speed in approach area is subject to controller's permission.

such 〔形〕そのような

航空機メーカーは、コンコードのような航空機の製造には消極的になるであろう。
Aircraft manufactures will take a negative attitude to build such aircraft as Concord.

suffix 〔可〕接尾辞

分類コードには、後ろに番号が添付されている。
The classification code is attached with suffix numbers.

suggest
〔他〕提案する
示唆する 暗示する
suggestion 〔可〕提案

パイロットは、風の変化を事前に知る手順を提案した。
The pilot suggested a procedure to detect windshift in advance.
天気図は、予定ルート上に寒気団が存在することを示唆している。
The weather chart suggests existence of a cold air mass on the planned route.

summarize〔他〕要約する
summary〔可〕概要 要約

要約は、複雑な事柄を理解することに効果がある。
The summary is effective to understand complicated matters.

super high frequency（**SHF**）
マイクロ波

マイクロ波は、レーダーや衛星通信によく適用されている。

S

	SHF is commonly applied to radar and satellite communication.
supersede 〔他〕～に取って代わる	双発ジェット機が4エンジンの大型機に取って代わりつつある。 Twin jets are superseding four engine large aircraft.
supply 〔不·可〕供給 〔他〕供給する	運航管理者は、飛行中のパイロットに必要な情報を提供する。 The dispatcher supplies pilots in flight with necessary information.
support 〔可·不〕支援 〔他〕支援する	飛行中のパイロットは、運航管理者によって支援されている。 Pilots in flight are supported by a dispatcher.
surface 〔可〕表面 〔形〕表面の 地上の 水上の	電波高度計は、地上と航空機の間の縦の距離を測定する。 The radio altimeter measures the vertical distance between the aircraft and the surface.
surveillance 〔不〕監視	国内航空交通は、レーダーで監視されている。 The domestic air traffic is kept under radar surveillance.
synchronize 〔自·他〕同調する 同調させる	目標物の方位は、アンテナの向きに同調している。 The azimuth of a target is synchronized with the antenna direction.
system 〔可〕体系 系統 組織網	航空機は、すべての系統の正常な作動によって安全に飛行する。 An aircraft flies safely relying on the normal function of all systems.

T

table 〔可〕表 テーブル	表は、進入から着陸までの解析データを示す。

The table presents analysis data of approach to landing.

---

tail
〔可〕尾部
〔形〕尾部の

引き起こしでピッチレートが大き過ぎると、尾部接触となる。
Excessive pitch rate at rotation causes a tail hit.

tailwind 〔不〕追い風

パイロットは、追い風では着陸しない。
Pilots do not land in tailwind.

---

takeoff 〔可·不〕離陸

パイロットは、離陸と着陸を繰り返し練習する。
A pilot repeatedly practices takeoff and landing.

takeoff clearance
離陸許可

離陸許可には、滑走路番号と風のデータが含まれる。
The runway number and the wind data are included in the takeoff clearance.

---

target
〔可〕目標物
〔他〕目標とする

管制官は、ターゲットがSA101であると確認した。
The controller identified the target as SA101.

---

taxi
〔可〕地上走行
〔自〕地上走行する

地上走行は、滑走路手前の停止線までに制限される。
Taxing is limited to the stop line proximate to the runway.

taxi clearance
地上走行許可

地上走行許可は、地上管制席が発行する。
Taxi clearance is issued by the ground control.

taxiway
タクシーウェイ誘導路

---

technical
〔形〕専門の 技術的

ATC通信の専門用語は、ICAOによって規定されている。
The technical terms of ATC communication are prescribed by ICAO.

technique〔不〕技法 技術

航空機は、最新技術の塊と言える。

T

technology〔不〕科学技術	An aircraft is regarded as a pack of advanced technique.
telecommunication〔不〕	電気通信
telegraphy〔不〕電信	モールス電信 Morse telegraphy
telephony〔不〕電話	無線電話 radio telephony
telephony designator 無線電話識別	各航空会社の無線局には、無線電話識別が与えられている。 A telephony designator is assigned to each airline radio station.
temperature 〔可·不〕温度 気温	国際標準大気における地表の温度は、15° Cである。 The surface temperature by ISA is 15° C.
outside air temperature 外気温度	機の上昇に伴って外気温度は低下する。 The outside air temperature decreases as the aircraft climbs.
temporary 〔形〕一時的な 臨時の	運航管理者は、降雪は一時的なものと予測した。 The dispatcher estimated it was a temporary snowfall.
temporarily 〔副〕一時的に 臨時に	表面の小さな凹みは、仮修理された。 A small dent on the skin is temporarily repaired.
term〔可〕用語 専門語	技術用語は、用語集にリストされている。 The technical terms are listed in the glossary.
terminal 〔可〕ターミナル 終点 端子 〔形〕終点の	飛行場の天候は、ぎりぎりと予報されている。 The terminal weather is forecasted to be marginal. 故障の原因は、端子の緩みかもしれない。 The trouble may be caused by a loose terminal.

terminal control area (TCA) ターミナル管制区	TCAではVFR機に対してもレーダーサービスが提供される。 Radar service is available for VFR aircraft in TCA.
terminate 〔自〕終る 〔他〕終える **termination** 〔不〕終了	受信局は、通報の終了前に遭難呼出しに対する受信証を送信してはならない。 A station should not send an acknowledgement of receipt for distress call before termination of the message.
terrain 〔可·不〕地形	地形に対する接近は、GPWSの音声で警告される。 Proximity to the terrain is verbally warned by GPWS.
test 〔可〕試験 検査 〔他〕試験する **test flight** 試験飛行	試験飛行によって、飛行性能及び各系統の機能が証明される。 Flight performance and system functions are proved by a test flight.
textbook 〔可〕教科書	学習者は、教科書に従ってATC英語を勉強すべきである。 Learners should study ATC English by the textbook.
therefore 〔副〕それゆえに	交通の流れは、スムースのようである。従って到着が遅れることはないであろう。 The traffic flow appears to be smooth. Therefore no arrival delay will occur.
threat 〔可〕脅迫、脅威 **bomb threat** 爆弾脅迫 爆破予告 **threaten** 〔自〕脅す ～が迫っている	嵐の脅威の下に under threat of an attack of storm 嵐が迫っている。A storm threatens.

T

through 〔前〕～を通って	VHF 及びそれ以上の周波数の電波は、電離層を貫通して衛星に到達する。 The radio waves of VHF and higher frequencies penetrate through the ionosphere and reach the satellite stations.
throughout 〔副〕～の間ずっと 至る所	旅客機は、隅から隅までしっかりと建造されている。 Airliners are well built throughout.
time 〔不〕時間 時刻	日本標準時は、世界協定時より9時間進んでいる。 The Japan standard time (JST) is 9 hours ahead of UTC.
tone 〔可〕 （音の）調子 気風 傾向	機がNDB上空を飛行する時ADFは、音程変調信号を発する。 ADF gives a tone modulated signal when the aircraft flies over the NDB station.
touch down 接地 接地する	機は、接地帯に接地した。
touchdown zone 接地帯	The aircraft touched down at the touchdown zone.
touchdown point 接地点	接地点は、前方へ伸び過ぎていた。 The touchdown point was far inside.
toward 〔前〕～に向かって	パイロットは、代替空港に向かって機を操舵した。 The pilot steered the aircraft toward the alternate airport.
tower 〔可〕タワー 塔	管制塔は、通常タワーと呼ばれる。 The control tower is usually called a tower.
traffic 〔不〕交通	交通の流れは、スムーズである。 Traffic flow is smooth.
traffic advisory（TA） 対抗機位置情報	TAは、パイロットに近づきつつある危険な対抗機について警告する。

T

TA warns a pilot of nearing hazardous traffic.

traffic alert and collision avoidance system (TCAS)
航空機衝突防止装置

TCASは、機上装備である。
TCAS is an airborne equipment.

traffic pattern 場周経路

管制官は、SA101に左場周経路の追風経路に合流するようにクリアランスを発した。
The controller issued clearance to intercept downwind leg of left traffic pattern to SA101.

---

train
〔可〕行列 連続 列車

多くの人が長い列を成して新機種の到着を待った。
There were many people of a long train waiting for the arrival of a new aircraft.

〔自･他〕訓練する
training 〔不〕訓練
training flight 訓練飛行

機種別のパイロット訓練コースは、短縮された。
The pilot training course of a type rating is shortened.

---

transfer 〔可･不〕移動

管制移管に伴って無線連絡が必要である。
Radio contact is required along with a control transfer.

〔自･他〕移す 乗換える

乗継ぎ客は、この空港で乗換える。
Transit passengers transfer at this airport.

---

transform〔他〕変圧する
transformer〔可〕変圧器

変圧器は、交流電圧を変化させる。
The transformer changes the AC voltage.

---

transit
〔不〕通過
〔形〕乗継ぎの

通過客は、乗継ぎ用ラウンジで待つ。
Thru passengers wait for departure at the transit lounge.

transition
〔可･不〕変遷 過渡期

パイロットは、新機種に乗務するには移行訓練を受けなければならない。
Transition training is required for pilots to be assigned on a new type aircraft.

transition route 転移経路	標準出発経路と航空路の間に、必要に応じ転移経路が設けられる。 A transition route is established between the standard departure course and an airway as required.
translate 〔他〕翻訳する **translation** 〔可·不〕翻訳	SELCALの解読器は、呼出装置を作動させるように受信信号を解読する。 The SELCAL decoder translates the received signal to activate the calling device.
transmit 〔他〕送る 伝導する 送信する **transmission** 〔可·不〕伝達 送信	電波は、遠くまで送信される。 Radio waves are transmitted to a far distance. 送信は、他の通信を妨害しないように注意深く行うべきである。 Transmission should be carefully made not to cause interference to the other communication.
transponder〔可〕応答機	応答機は、質問機に応答信号で返答する。 The transponder responds the interrogator with a reply signal.
transport 〔他〕運送する 〔可〕輸送機　〔不〕輸送	航空会社は、乗客及び貨物を輸送する。 Airlines transport passengers and cargo.
travel 〔自〕旅行する 伝わる	直接波は、送信アンテナから受信アンテナへ直接伝わる。 Direct waves travel directly from a transmitting antenna to a receiving antenna.
troposphere 〔不〕対流圏	対流圏では、電波の伝播は乱れる。 Radio waves' propagation is unstable in the troposphere.
turbine 〔可〕タービン	ジェットエンジンのタービン温度は、運用上非常に重要な要素である。

T

Turbine temperature of a jet engine is a critical factor in operation.

---

turbulence
〔不〕（大気）の乱れ
turbulent 〔形〕荒れた

パイロットは、擾乱の場所について知らなかった。
The pilot was unaware of the turbulent air area.

---

turn 〔自·他〕
回る 回す（方向を変える）

機は、着陸復行後180°旋回をした。
The aircraft made 180° turn after executing go-around.

---

twice 〔副〕二度 二回

機は、しっかりと接地する前に二度バウンドした。
The aircraft bounded twice before firm touchdown.

---

type 〔可〕型式

旅客機は、型式Tの航空機に分類されている。
Airliners are categorized in T-type aircraft.

type of approach
進入方式

ILSまたは目視進入は最も普通の進入方式である。
ILS approach or the visual approach is the most common type of approach.

---

typical 〔形〕代表的な

VOR/DME情報を基にしたRNAVルートは、代表的な国内航空路である。
RNAV routes referring to the VOR/DME information are the typical domestic routes.

U

ultra high frequency (UHF)
極超短波

グライドパスは、UHFで作動する。
Glide path operates on UHF.

---

unable
〔形〕～ができないで

局と連絡が出来なかった場合には、航路上の他の周波数で連絡を試みること。
When you are unable to contact a

station, try to contact with other frequency appropriate to the route.

---

uncertain 〔形〕不確実な

低高度における突風の存在は、はっきりしない。

Existence of a low level windshear is uncertain.

uncertainty 〔不〕不確実

昔の飛行は、予測の付かないことだらけであったと言われている。

It is said that an old days' flight was full of uncertainties.

---

under 〔前〕～のもとに

離陸または着陸の航空機は、タワーの管制下にある。

The aircraft taking off or landing is under the control of the tower.

---

understand 〔自·他〕理解する

管制官の指示を正確に理解することが重要である。

It is important to correctly understand the controller's instruction.

---

unit 〔可〕機関

気象観測所は、気象庁の機関である。

The weather observatory is a unit of the Meteorological Agency.

---

unlawful 〔形〕不法の

所定の無線機器を装備しないで走行区域内で車両を運用することは、違法である。

It is unlawful to operate an automobile on the maneuvering area without specified radio equipment.

---

unless
〔接〕～でない限り
〔前〕～を除いては

機長は、燃料が厳しくならない限り天候回復を待つために待機すると決めた。

The captain decided to hold for weather recovery unless fuel becomes short.

---

unreadable

そちらの通信は、聞き取れない。

〔形〕聞き取れない	Your message is unreadable.
until〔前〕～まで	パイロットは、霧が晴れるまで待機を継続すると決めた。 The pilot decided to continue holding until the fog clears.
update〔他〕最新化する	システムの改修が行われると、機の操作手順が最新のものにされる。 Aircraft operating procedures are updated when a system is modified.
upper〔形〕より高い方の	上空の温度の方が、地表の温度より低い。 The upper air temperature is lower than the surface temperature.
upper limit〔可〕上限	使用できる高度の上限は、41,000フィートである。 The upper limit of the usable altitude is 41,000ft.
urgency〔不〕緊急	エンジン火災は、重大な緊急の問題である。 Engine fire is a problem of great urgency.
urgency condition 緊急事態	緊急事態は、トランスポンダーコード7700をセットすることにより知らせられる。 The urgency condition is reported by setting the transponder code 7700.
urgent〔形〕切迫した	その人は、緊急医療を必要とする。 The man is in urgent need of medical assistance.
use	
〔不〕使用 利用	スピードブレーキの使用は、減速に有効である。 Use of the speed brake is effective for deceleration.
〔他〕使用する	レーダーには、UHFがよく使われる。

U

UHF is often used for radar.

useful 〔形〕有用な

運航管理者は、飛行中のパイロットに有用な情報を提供する。

The flight operations officer provides pilots in flight with useful information.

usual 〔形〕通常の

何時もの好天気で、パイロットは降下着陸での困難な作業から免れた。

The usual fine weather relieved pilots from hard work during descent and landing.

usually 〔副〕通常

この季節の使用滑走路は、通常34Rである。

The active runway is usually 34R in this season.

V

vary

〔自〕変わる 変更する
〔他〕変える 変化する

停止距離は、滑走路の摩擦係数によって異なる。

The stopping distance varies in relation to a coefficient of the runway friction.

variable 〔形〕可変の

可変ピッチプロペラは、効率的な推進装置である。

The variable pitch propeller is efficient propulsion equipment.

variable phase 可変位相

VORは、基準位相と可変位相の差から方位を決める。

VOR determines the bearing from the difference between the reference phase signal and the variable phase signal.

variation 〔可・不〕変動

周波数の変動は、周波数カウンターで検知する。

Frequency variation is detected with a frequency counter.

various 〔形〕さまざまな

飛行中の燃料節減方法に関するアイディアは、いろいろ沢山ある。

The ideas of in-flight fuel saving

methods are many and various.

---

vector
〔可〕方向
〔他〕（電波によって）誘導する

PARの管制官は、機を最終進入開始地点へと誘導した。
PAR controller vectored the aircraft to the final approach initiation fix.

---

vehicle 〔可〕乗り物

走行区域で運用される乗り物は、登録されている。
The vehicles operated on the maneuvering area are registered.

---

verbal〔形〕口頭の 言葉の

口頭での報告で十分である。
A verbal report is good.

verbally
〔副〕口頭で 言葉で

口頭または文書で回答しなさい。
Reply verbally or in writing.

---

verify 〔他〕確かめる

パイロットは、天候が最低気象条件を満たしていることを確認した。
The pilot verified the weather condition met the minimum requirement.

verification 〔不〕確認

パイロットは、フラップを出す前に速度の確認をしなかった。
The pilot lacked verification of IAS before extending flaps.

---

version 〔可〕版

最新版 the latest version

---

vertical 〔形〕縦の

航空路上の縦の間隔は、通常2,000フィートである。
The vertical separation between aircraft on a airway is normally 2,000ft.

VHF omni-directional radio range （**VOR**）
超短波全方向式無線標識

国内航空路は、VOR局を結んで形成されている。
Domestic airways are formed by connecting VOR stations.

---

via 〔前〕～経由で

この便は、ホノルル経由ロスアンジェルス行きである。

This flight is for Los Angeles via Honolulu.

---

vicinity 〔可〕周辺 付近

パイロットは、滑走路へと進入する時、病院の上空及びその周辺の飛行を避ける。

Pilots avoid flying over the hospital and its vicinities while they approach the runway.

---

video
〔不〕ビデオ 〔形〕映像の

レーダーの受信装置は、受信信号を音声信号、映像信号またはデジタルデータに変換する。

The receiver section of a radar system converts received signal into the audio signal, video signal or digital data.

---

view 〔可･不〕眺望 見解

viewpoint 〔可〕観点

マニュアルでは、航空機の各系統は操作の観点に立って説明されている。

Aircraft system is explained from an operational viewpoint in the manual.

---

visibility 〔不〕視程

実際の視程は、予報よりずっと良かった。

The actual visibility was much better than it had been forecasted.

---

visual 〔形〕視覚に関する

進入経路上に幾つかの目視キューがあると、自機の飛行経路のコントロールが容易になる。

Having some visual cues along the approach course facilitates path management.

visual flight rule (**VFR**)
有視界飛行方式

レーダーサービスは、もし要求があればVFR機にも提供される。

Radar service is provided to a VFR aircraft if it is requested.

---

voice 〔可･不〕音声

マイクロフォンは、音声を電気信号に

	変換する。 A microphone converts voice into electric signals.
voice calling 音声による呼出し	SELCALは、音声呼出しに取って代わった。 SELCAL has superseded voice calling.
voice communication 音声通信	ATCは、管制官とパイロットの音声通信を基本としている。 ATC is based on the voice communication of a controller and a pilot.
voltage 〔可·不〕電圧	無線機器は、低電圧で作動する。 A radio apparatus operates on low voltage.
volume 〔可·不〕容量 音量	タンク内の燃料量は、重量で測られる。 The volume of fuel in the tank is measured by weight.

W

wait 〔自·他〕待つ	パイロットは、霧が晴れるまで待った。 The pilot waited till the fog cleared.
warn 〔他〕警告する	RAは、パイロットに即座の回避行動をとるように警告する。 RA warns pilots to take immediate avoidance action.
warning 〔可·不〕警告	パイロットは、SIGMETは悪天候の警告と受け止める。 Pilots regard SIGMET as a warning of severe weather conditions.
watch 〔不〕見張り 〔他〕監視する	通信のウォッチ（聴守）は、パイロットの義務である。 Pilots are obliged to watch radio communication.

waveguide 〔可〕導波管	導波管は、マイクロウェーブに適用されている。 A waveguide is applied to microwaves.
weather 〔不〕気象 天気	好天は、乗客に快適な飛行をもたらす。 Favorable weather contributes a comfortable flight to passengers.
weather information 気象情報	気象情報を含む飛行場の状態は、ATISで放送されている。 Airport conditions including weather information are broadcasted with ATIS.
weather minima 最低気象条件	気象状態が最低気象条件を満たさない限り、離陸許可は発行されない。 Takeoff clearance is not issued unless the weather condition meets the weather minima.
weather radar 気象レーダー	種々の飛行データや航法データが、気象レーダー画面に表示される。 Various flight data and navigation data are displayed on the weather radar screen.
weather report 気象報告	パイロットは、悪天候を観察した場合には気象報告を行うように要求されている。 Pilots are requested to make a weather report when they have observed a significant weather condition.
whatever 〔代〕いかに～であっても	パイロットは、事態が何であろうと安全に着陸しなければならない。 A pilot must land safely whatever the situation is.
whole 〔形〕 ちょうど 全体の すべての	ちょうど百の数 the number of the whole hundred

W

wide 〔形〕幅の広い	レーダーは、幅広くターゲットを捜索する。 Radar searches targets for a wide area.
wide range 広範囲の	運航に携わるには、広範囲の知識が必要である。 A wide range of knowledge is required to be engaged in flight operations.
width 〔可·不〕広さ 幅	レーダーの方位決定能力は、ビーム幅と関連している。 The azimuth resolution capability of radar is related to the beam width.
wind 〔不〕風	離陸のクリアランスには、風の情報が含まれる。 Wind data is included in the takeoff clearance.
wind direction 風向	風向は、変化しつつある。 Wind direction is shifting.
wind velocity 風速	風速は不安定で、突風性である。 Wind velocity is unstable and it is gusty.
windshear 突風	パイロットは、低高度の突風に気をつける。 A pilot is careful about a low level windshear.
windshield 風防	古風の飛行機では、操縦席の前に風防がある。 An old fashion airplane has a windshield in front of the pilot's seat.
windshift 風の変化	低高度での急な風の変化は、機の安定を乱す。 Rapid windshift at low altitude interrupts aircraft stability.
wing 〔可〕翼	主翼の面積及び揚力係数は、フラップにより制御される。

wing leading edge 翼の前縁	The area and the lift coefficient of the main wing are controlled by flaps.
wish 〔自·他〕望む	乗客は、快適な飛行を望む。 Passengers wish an enjoyable flight.
with 〔前〕～とともに	VOR局には、DMEも装備されている。 VOR station is equipped with DME.
without 〔前〕～なしに	DMEのないVOR局はない。 There is no VOR station without DME.
within 〔前〕～以内で	レーダの覆域以内 within the radar coverage
word 〔可〕語 単語	通報の各語は、はっきりと発音されるべきである。 Each word of a message should be clearly pronounced.
world 〔可〕世界 **worldwide** 〔形〕世界的な	世界規模のATC英語の標準化が重要である。 Worldwide standardization of ATC English is a matter of importance.
write 〔自·他〕書く	読むことと書くことは、外国語の理解を上達させる。 Writing and reading improves understanding of a foreign language.

Z

zone 〔可〕地域 区域	騒音軽減方式は、住宅地区の騒音軽減を目指している。 Noise abatement procedures aim to reduce noise over the residential zone.

第2章

和英辞書

凡　例

〔動〕：動詞
〔名〕：名詞
〔代〕：代名詞
〔助〕：助動詞
〔形〕：形容詞
〔副〕：副詞
〔前〕：前置詞
〔接〕：接続詞

あ	
合図〔名〕	signal, sign
合図する〔動〕	signal, give a signal, make a signal
合図の〔形〕	signal light, signal flag
間〔名〕	interval (間隔) opening (隙間) for (for an hour)
～の間ずっと〔前〕	during (during cruise 巡航中)
～する間〔接〕	while (while turning 旋回中)
間の 間に 間で ～の中で〔前〕	between (between cloud layers 雲の層の間で) among (among the crowd 人込みの中で)
明らかな〔形〕	clear, obvious, apparent, distinct, evident clear sky (晴天) evident mistake (明らかな間違い)
明らかに はっきりと〔副〕	clearly pronounce (はっきり発音する)
悪天候〔名〕	significant weather, bad weather, rough weather
悪天情報〔名〕	significant meteorological information (SIGMET)
浅い〔形〕	shallow (a shallow descent path 浅い降下経路)
与える〔動〕	give, provide, supply, furnish, render, supply fuel (燃料を補給する)
宛名 宛先〔名〕	address (address of the message 通報の宛先)
アプローチ 進入〔名〕	approach (final approach course 最終進入コース)
あらまし 概要〔名〕	outline (outline of the weather 天候の概要) summary (summary of the analysis 解析の概要)
概説する 要約する〔動〕	outline (outline the operating procedure 操作手順を概説する) summarize (要約する)

ありそうな〔形〕	likely (It's likely to rain. 雨が降りそうである。)
アルファベット〔名〕 アルファベットの〔形〕	alphabet alphabetical (alphabetical order アルファベット順)
暗号 符号〔名〕 暗号化する 符号化する〔動〕 暗号を解く〔動〕	code, cipher (暗号) code, encode decode
安全〔名〕 安全な〔形〕 安全に〔副〕	safety (flight safety 飛行の安全) security (public security 治安、公安) safe (safe landing 安全な着陸) safely (land safely 安全に着陸する)
安定〔名〕 安定する 安定させる〔動〕 安定した〔形〕	stability (speed stability 速度の安定) stabilize (stabilize the altitude 高度を安定させる) steady (steady wind 安定した風)
アンテナ〔名〕	antenna (antenna element アンテナ素子)
い	
以下の〔前〕	below (below sea level 海面下の) (the table below 下記の表)
いくらかの〔形〕	some (There are some birds on the runway. 滑走路上に何羽かの鳥がいる。)
移行 移転〔名〕 移行する〔動〕	transfer (control transfer 管制移管) transfer (transfer communication to the next control 通信を次の管制に移行する)
維持する〔動〕	maintain, hold, keep, stay, retain, sustain(maintain present speed 現在の速度を維持する)
以前の〔形〕 以前に(は)〔副〕	previous (previous message 以前の通報) previously (previously issued clearance 以前に発行されたクリアランス)

忙しい〔形〕	busy, occupied (busy with overlapped work 重複作業で忙しい)
依存する 頼る〔動〕	depend (Flight safety depends on ATC. 飛行の安全は、ATCに依存する。)
位置〔名〕	position, place, spot, location, site (present position 現在位置)
位置する〔動〕	locate (a satellite located over the equator 赤道上空に位置する衛星)
一時的な〔形〕	momentary (momentary confusion 一時的混乱)
一時的に〔副〕	momentarily (The aircraft sank momentarily. 機は、一瞬沈んだ。)
一時的休止 ポーズ〔名〕	pause (a pause between words 語の間のポーズ)
一次の〔形〕	primary (primary radar 一次レーダー)
著しい〔形〕	significant, meaningful (significant change in temperature 温度の著しい変化)
一度〔副〕	once (The aircraft once bounded at landing. 機は、着陸時に一度バウンドした。)
一括呼出し〔名〕	all call (The controller sent the message by all call. 管制官は、通報を一括呼出しで送信した。)
一致 同意 調和 一貫性〔名〕	agreement, accord, consistency consensus (意見の一致)
一致する〔動〕	agree, accord (The forecast weather data agree with the actual condition. 予報気象データは、実情と一致する。)
一片 部分 断片〔名〕	piece (一片) section (部分) slice (断片) (a piece of broken panel 破損したパネルの一片)

逸脱 偏差〔名〕	deviation (standard deviation 標準偏差)
逸脱する 外れる〔動〕	deviate (deviate from the course コースから外れる)
意図 意向〔名〕	intention (pilot's intention パイロットの意向)
意図する〔動〕	intend (意図する) plan (計画する) propose (企てる) aim (めざす) (The pilot proposed the optimum altitude. パイロットは、最適高度の飛行を企てた。)
緯度〔名〕 経度〔名〕	latitude (緯度) longitude (経度) (The fix is indicated by longitude and latitude. その地点は、緯度、経 度で表示される。)
移動業務〔名〕	mobile service (aeronautical mobile service 航空移動業務)
以内の〔前〕	within (within Fukuoka FIR 福岡FIR内の)
意味〔名〕	meaning, meaningfulness
意味する〔動〕	mean (SIGMET means dangerous weather. SIGMETは、危険な気象を 意味する。)
意味ある 意味深い〔形〕	meaning, meaningful
意味ありげに 意味深く〔副〕	meaningfully
以来〔接〕	since (It has been a long time since the first jet airliner was introduced. 最初のジェット旅客機が導入されて以 来長い年月が経っている。)
医療の〔形〕	medical (medical care 医療)
インピーダンス〔名〕	impedance (impedance matching インピーダンス整合)
う	
ウィンドシア〔名〕	windshear (low level windshear 低高度ウィンドシア 低高度突風)

ウェイブガイド 導波管〔名〕	waveguide (waveguide for microwaves マイクロ波用のウェイブガイド)
受入れる 受け取る 許す〔動〕	accept (受入れる) receive (受取る) admit (許す) (The controller accepted pilot's request. 管制官は、パイロットの要求を受け入れた。)
失う〔動〕	lose (lose sight by glare まぶしい光で視界を失う)
疑い 疑問〔名〕	doubt, question
疑う〔動〕	doubt
疑わしい〔形〕	doubtful, questionable (a questionable forecast 疑わしい予測)
海の 海事の〔形〕	marine, maritime (maritime museum 海事博物館)
雲高〔名〕	ceiling, cloud height (Ceiling is the distance between the cloud base and the surface. 雲高は、雲底と地表の間の距離である。)
雲底〔名〕	cloud base
運航〔名〕	flight operation
運航管理〔名〕	flight operations control
運航管理者〔名〕	flight operations officer, dispatcher
運航管理通信〔名〕	aeronautical operations control communication (AOC)
運航業務通信〔名〕	aeronautical administrative communication (AAC)
運転 駆動装置〔名〕	drive (front wheel drive 前輪駆動)
運転する 駆動する〔動〕	drive (engine driven generator エンジンにより駆動される発電機)
運輸 輸送〔名〕	transport, transportation
航空輸送〔名〕	air transportation
輸送機〔名〕	transport airplane airliner (定期旅客機)

運輸多目的衛星〔名〕	multi-functional transport satellite (MTSAT)

え

影響 効果〔名〕	effect
影響する〔動〕	affect (Strong headwind affects the arrival time. 強い向かい風は、到着時刻に影響する。)
衛星〔名〕	satellite
衛星の軌道〔名〕	satellite orbit
衛星航法〔名〕	satellite navigation
衛星通信〔名〕	satellite communication
映像の〔形〕	video (video data 画像データ)
沿岸〔名〕	coast (the coast of Japan 日本の沿岸)
沿岸の〔形〕	coastal (coastal area 沿岸地域)
援助 支援〔名〕	assist, support, aid, assistance (technical assistance 技術援助)
援助する 支援する〔動〕	aid (navigation aid facility 航法援助施設)
エンジン〔名〕	engine (four engine jet-aircraft ４発ジェット機)
円柱状のもの 欄〔名〕	column (control column 操縦桿)

お

追風〔名〕	tailwind (tailwind component 追風成分)
応じる 従う〔動〕	comply (comply with the controller's instruction 管制官の指示に従う)
応諾 応じること〔名〕	compliance (in compliance with the clearance クリアランスに従って)
大きい〔形〕	great (great circle course 大圏コース)
送り込む 供給する〔動〕	feed (feed line 給電線)
送る〔動〕	send (send a message　通報を発信する)

音〔名〕	sound (speed of sound 音速)
音を出す〔動〕	sound (The SELCAL chime sounded. SELCALのチャイムが鳴った。)
音の調子 音色〔名〕	tone (a high tone 音の高い調子) (tone modulation 可聴音による変調)
衰える〔動〕	attenuate (The signal strength of the ground surface waves attenuate during propagation. 地上波の信号は、伝播中に減衰する。) fade (The radio signal strength faded out. 無線信号は、次第に衰えていった。)
各々の〔形〕	each (A call sign is assigned to each flight. 各フライトに、コールサインが与えられる。)
終り〔副〕	over (送信の終りを意味するATC用語)
終る 完了する〔動〕	end, terminate, finish, complete
音声〔名〕	voice (voice communication 音声通信) (voice calling 音声呼出し)
音速〔名〕	speed of sound
温度〔名〕	temperature (outside air temperature 外気温度)

か

開始〔名〕	beginning, start, initiation, commencement
開始する〔動〕	commence (commence takeoff 離陸を開始する) begin, start, initiate, launch
会社〔名〕	company
海事の〔形〕	maritime (maritime law 海事法)
回折〔名〕	diffraction (diffraction waves 回折波)

回折する〔動〕	diffract (Radio waves are diffracted near the surface. 電波は、地表近くでは回折する。)
介入 干渉〔名〕	intervention, interference (unlawful interference 不法介入)
介入する 妨げる 干渉する〔動〕	intervene, interfere
回避〔名〕	avoidance
回避する〔動〕	avoid (avoid bad weather 悪天候を避ける)
回避情報〔名〕	resolution advisory (RA) (TCASの回避情報)
回復〔名〕	recovery (recovery of normal operation 通常操作の回復) restoration (restoration of order 秩序の回復)
回復する〔動〕	recover, restore (restore normal condition 通常状態を回復する)
海里〔名〕	nautical mile
回廊〔名〕	corridor (air corridor 空の回廊)
会話〔名〕	conversation
会話する〔動〕	talk (話す)
係り 係り合い〔名〕	relation, connection
係る 係り合う〔動〕	involve, concern
～に拘らず〔接〕	though, although
～に拘わらず〔前〕	despite (Flight condition is smooth despite being in cloud. 雲の中にも拘わらず飛行状態はスムーズである。)
核心 根源〔名〕	core (core satellite 核心をなす衛星) center, root, focus
確信〔名〕	certainty (certainty of safe flight 安全な飛行への確信)

か

拡大 延長 増加〔名〕	extension (runway centerline extension 滑走路中心線の延長) expansion (expansion of the airport 空港の拡張) increase (increase in cost 経費の増加)
拡大する 延長する 増加する〔動〕	extend, expand, increase
確認〔名〕	confirmation, verification, identification
確認する〔動〕	confirm (confirm the assigned altitude 指定された高度を確認する) verify (verify the airspeed 速度を確認する) identify (identify by a name tag 名前札で確認する)
確保する 確実にする〔動〕	secure (secure the seat 席を確保する) ensure, assure
火災警報〔名〕	fire warning (Takeoff was aborted because the fire warning system was activated. 火災警報装置が作動したので、離陸は中断された。) fire alarm (火災警報)
風〔名〕	wind (風) breeze (微風) draft (すきま風) headwind (向い風) tailwind (追い風) crosswind (横風) gust (突風) monsoon (季節風) wind data (風のデータ) crosswind component (横風成分)
ウィンドシア (突風)〔名〕	windshear (Low level windshear warning system is under development. 低高度ウィンドシア警報装置は、開発中である。)
型式 種類〔名〕	type (an aircraft of a new type 新型式の航空機) kind, sort

形〔名〕	form (a report in the specified form 所定の形式の報告) format（体裁） shape (形)
滑走路〔名〕	runway (runway length 滑走路の長さ)
活動〔名〕	activity (The volcano is in activity. 火山は、活動している。)
活動的な〔形〕	active (active volcano 活火山)
過渡期〔名〕	transition (transition period from subsonic speed to supersonic speed 亜音速から超音速への過渡期)
可能性〔名〕	possibility (a remote possibility まれな可能性)
可能な〔形〕	possible (possible but difficult 可能だが難しい)
可変的な〔形〕	variable (variable wing 可変翼)
かまわず〔形〕〔副〕	regardless (regardless of bad weather 天気が悪いにもかまわず)
雷〔名〕	thunder (雷鳴) lightning (稲光) thunderstorm (雷雨 雷電)
加盟国〔名〕	member state, member country
仮の〔形〕	temporary (temporary repair 仮の修理)
仮に〔副〕	temporarily (It's repaired temporarily. それは仮に修理された。)
変わる 変える〔動〕	change (The weather will change. 天気は、変わるだろう。) shift (The wind will shift. 風は、変わるだろう。) vary (The weather varies hourly. 天気は、刻々と変化する。)
間隔〔名〕	interval (an interval between pulses パルスとパルスの間の間隔)

	separation (a separation between aircraft 航空機間の間隔)
環境〔名〕	environment (the environment 自然環境)
環境の〔形〕	environmental (environmental pollution 環境汚染)
関係〔名〕	relation (the relation between thrust and drag 推力と抗力との関係) relationship (関係 関連)
関係する〔動〕	relate (The ground speed relates with the wind component. 対地速度は、風の成分に関係する。)
簡潔な〔形〕	concise (a concise message 簡潔な通報)
簡潔に〔副〕	concisely (state concisely 簡潔に述べる)
刊行物〔名〕	publication (AIP is an official publication. AIPは、公の刊行物である。)
観察 遵守〔名〕	observation (make observation of the sun 太陽の観察をする) watch (keep watch on the emergency frequency 非常用周波数を監視する) observance
観察する 監視する〔動〕	observe, watch
干渉〔名〕	interference (harmful interference 有害な混信)
干渉する 邪魔をする〔動〕	interfere
関心 関心事〔名〕	concern interest (take interest in ～に関心を持つ)
完成 完了〔名〕	completion (completion of flight 飛行の完了)
完成する 完了する〔動〕	complete (complete a turn 旋回を完了する)
管制〔名〕	control (air traffic control 航空交通管制)
管制する〔動〕	control (control the separation 間隔を整える)

管制移管〔名〕	control transfer
管制官〔名〕	controller
管制区域〔名〕	control area
管制空域〔名〕	controlled air space
管制圏〔名〕	control zone
管制許可〔名〕	clearance
管制許可伝達席〔名〕	delivery section
管制席〔名〕	control section
管制塔〔名〕	control tower, tower
慣性航法装置〔名〕	inertial navigation system (INS)
慣性基準装置〔名〕	inertial reference system (IRS)
完全な 素晴しい〔形〕	perfect (a perfect flight 素晴しいフライト)
完全に〔副〕	perfectly
簡単な 短い 容易な〔形〕	simple, brief, easy
簡単に 手短に 容易に〔副〕	simply, briefly, easily
貫通 浸透〔名〕	penetration
貫通する 浸透する〔動〕	penetrate (VHF and higher frequency radio waves penetrate the ionosphere. VHF及びそれ以上高い周波数の電波は、電離層を貫通する。)
観点〔名〕	point of view, viewpoint, standpoint
慣用句 語 熟語〔名〕	idiom, phrase
関連〔名〕	relation, relationship, connection in connection with (〜に関連して) in relation to (〜に関連して)
関連する〔動〕	relate
該当する 適用できる〔形〕	applicable (The fuel saving procedure is applicable to a long range flight. 燃料節減方式は、長距離飛行に適用できる。)
概念 構想〔名〕	concept (The concept of flight opera-

か

	tion is safety. 運航の構想は、安全である。)
外部の 外の〔形〕	external (external inspection 外部点検) outside (outside watch 外の監視)
画面 スクリーン〔名〕	screen (radar screen レーダーのスクリーン)
き	
機関 組織体〔名〕	unit (a unit of air traffic control 航空交通管制の機関) organization (組織体)
単位〔名〕	unit (The feet is a unit of length. フィートは、長さの単位である。)
機関 エンジン〔名〕	engine (an internal-combustion engine 内燃機関)
聞く 聞こえる〔動〕	listen, hear
危険〔名〕 危険な〔形〕	danger, risk, hazard dangerous, hazardous, risky, unsafe, grave
起源 発端 原因〔名〕	origin (origin of communication 発信元)
起源の〔形〕	original (the original plan 原案)
帰航の 入ってくる〔形〕 出航の〔形〕	inbound (inbound flight 帰航便) outbound (outbound flight 出航便)
機首方位〔名〕	heading (change heading 変針)
気象〔名〕	weather (weather radar 気象レーダー)
気象の〔形〕	meteorological (Meteorological Agency 気象庁)
記述 説明書〔名〕 記述する 説明する〔動〕	description describe (The noise abatement procedure is described in the manual. 騒音軽減方式は、規定に記述されている。)

基準 標準〔名〕	reference (reference point of an airport 飛行場標点)
機上の〔形〕	airborne (airborne equipment 機上の装備)
機上衝突防止装置〔名〕	airborne collision avoidance system (ACAS)
機上気象報告〔名〕	pilot report (PIREP)
機上通報〔名〕	air report
規制する 調整する〔動〕	regulate, control, adjust, coordinate
規則 規定 法規〔名〕	regulation, rule
基礎 根拠〔名〕	base, foundation
基礎的 基本的〔形〕	basic (basic operating weight 基礎運航重量) fundamental (fundamental factor 基本要因)
基礎を置く〔動〕	base on (～に～の基礎を置く)
基地局〔名〕	base station
機長〔名〕	pilot in command (PIC)
～に気づいて ～を知って〔形〕	be aware of (be well aware of ～を熟知して)
規定 規則 法規 法律 法令〔名〕	regulation, law, rule, prescript (法令)
処方箋〔名〕	prescription (処方箋)
規定する 処方する〔動〕	prescribe (規定する 処方する) stipulate (規定する)
軌道〔名〕	orbit
軌道に乗せる 軌道に乗って回る〔動〕	orbit (The spacecraft is orbiting around Mars. 宇宙船は、火星の周りを軌道に乗って回っている。)
気にする 心配する〔動〕	care, concern (The pilot was concerned about the windshift. パイロットは、風の変化を気にしていた。)

機能〔名〕 機能上の〔形〕	function functional (functional trouble 機能上の故障)
厳しい 厳格な 真剣な〔形〕	severe, strict, hard, serious
救出〔名〕 救出する 救助する〔動〕	rescue rescue, save (救助する)
給電線〔名〕	feeder, feed line (The coaxial cable is widely used as a feeder. 同軸ケーブルは、給電線として広く使用されている。)
急な 速い 迅速な〔形〕 急に 速く 迅速に〔副〕	rapid, quick, fast, swift, speedy rapidly, quickly, fast, swift
救難調整本部〔名〕	rescue coordination center (RCC)
救命無線機〔名〕	emergency locator transmitter (ELT)
教科書〔名〕	textbook
供給〔名〕 供給する 提供する 与える〔動〕	supply, provision supply, provide, give, furnish (ILS provides approach guidance by means of radio signals. ILSは、電波による進入時の誘導を提供する。)
共振 共鳴 反響〔名〕 共振する 共鳴する 反響する〔動〕	resonance resonate (An antenna resonates with an appropriate frequency radio wave. アンテナは、適切な周波数の電波と共振する。)
強調〔名〕 強調する〔動〕	emphasis (repeat the word for emphasis 強調のために語を繰返す) emphasize
共通の〔形〕	common (Common understanding of a pilot and the controller is essential for safe flight. パイロットと管制官の共通の理解が、安全な飛行のために必

	須である。)
一般に〔副〕	commonly (Microwaves are commonly applied to a radar system. マイクロ波は、一般にレーダーに使われている。)
脅迫 おどし〔名〕	threat
脅迫する おどす 差し迫る〔動〕	threaten, impend (差し迫る) (impending hot start 危険が差し迫ったホットスタート)
強烈な 極端な 激しい 強い〔形〕	intense, intensive, strong, violent
強烈に〔副〕	intensively, violently, strongly
許可 承認〔名〕	permission, approval, authorization
許可する 承認する〔動〕	permit, approve, authorize
拒絶 拒否 排除 辞退〔名〕	rejection, refusal
拒絶する 拒否する 辞退する〔動〕	reject, refuse (The pilot rejected takeoff. パイロットは、離陸を断念した。)
距離〔名〕	distance
距離線〔名〕	distance line (Intersection of three distance lines between an aircraft and satellites is the position of the aircraft. 航空機と衛星の間の三本の距離線の交差点は、機の位置である。)
距離測定装置〔名〕	distance measuring equipment (DME)
規律〔名〕	discipline (cockpit discipline 操縦室における規律)
記録 記号 マーク〔名〕	record, mark
記録する〔動〕	record (The pilot recorded the flight data in the logbook. パイロットは、ログブックに飛行データを記録した。)

き

緊急〔名〕	emergency, urgency
緊急の〔形〕	urgent, emergency (emergency descent 緊急降下)
緊急に〔副〕	urgently
緊急状態〔名〕	emergency condition, urgency condition
擬似の 偽りの〔形〕	pseudo
擬似局〔名〕	pseudo station
技術 技法〔名〕	technique
技術の 専門の〔形〕	technical (technical term 専門語)
義務〔名〕	obligation , duty
義務的な 必須の〔形〕	compulsory (compulsory aircraft station 義務航空機局) obligatory (obligatory subject 必修科目)
義務を負わせる〔動〕	oblige
行政 管理〔名〕	administration (行政 管理)
行政上の 管理上の〔形〕	administrative (administrative district 行政区画)

く

空域〔名〕	air space
空域管制〔名〕	area control (area control center (ACC))
空間 宇宙〔名〕	space (space development 宇宙開発)
空港 飛行場〔名〕	airport, aerodrome
空港面探知レーダー〔名〕	airport surface detection equipment (ASDE)
区分 部分 部門〔名〕	segment, division, category (Radio waves are divided into eight categories. 電波は、八つの部分に区分されている。)
区分する〔動〕	categorize, divide
組合せ 結合〔名〕	combination (Combination of a propeller and a turbo jet is called a

	turbo-prop. プロペラとターボジェットの組合せは、ターボプロップと呼ばれる。)
組み合わせる 複合する〔動〕	combine (MTSAT is a combined satellite of flight mission and meteorological mission. MTSATは、フライトミッションと気象ミッションを複合した衛星である。)
組入れる 合併する〔動〕	incorporate
雲〔名〕	cloud (cloud top 雲の最上点 cloud base 雲の基部)
シーリング 雲高〔名〕	ceiling (Ceiling is the distance between the surface and the cloud base. シーリングは、地表面と雲底の間の距離である。)
繰返し 反復 復唱〔名〕	repetition, read-back
繰返す 反復する 復唱する〔動〕	repeat, read-back (repeat for confirmation 確認のために繰返す)
訓練 養成 練習〔名〕	training, drill, practice, exercise
訓練する 養成する〔動〕	train, drill
訓練飛行〔名〕	training flight
グライドスロープ〔名〕	glide slope (Normal glide slope is inclined three degrees to the horizontal line. 通常のグライドスロープは、水平線に対して三度傾斜している。)
グライドパス〔名〕	glide path (An aircraft descend on the glide path. 航空機は、グライドパスに沿って降下する。)

け

経過 過程〔名〕	process, progress
経過する〔動〕	elapse (elapsed time 経過時間)
警戒 警告 警報〔名〕	alert, warning, alarm
警告する 注意する〔動〕	warn
計画 案〔名〕	plan (flight plan 飛行計画)

け

計画する 立案する〔動〕	plan (The pilot planned a high speed cruise. パイロットは、高速巡航を計画した。)
計器進入〔名〕	instrument approach
計器着陸方式〔名〕	instrument landing system (ILS)
計器飛行状態〔名〕	instrument meteorological condition (IMC)
計器飛行方式〔名〕	instrument flight rule (IFR)
警告 注意〔名〕 警告する 注意する〔動〕	warning, alarm, alert, caution warn, alert, alarm, caution
経済的な〔形〕 経済的に〔副〕	economical economically
傾斜 斜面 勾配〔名〕 傾斜した〔形〕 傾斜させる〔動〕	slant, slope, incline slant (slant distance by DME DMEによる傾斜距離) incline (The glide slope is inclined approximately 3° to the horizontal line. グライドスロープは、水平線に対して約3度傾斜している。)
継続 持続〔名〕 継続する 持続する 残る〔動〕	continuance, continuation, maintenance continue, maintain, last
形態〔名〕	mode (receiving mode 受信モード)
系統〔名〕	system (navigation system 航法系統)
経度〔名〕 経度の 縦の〔形〕	longitude longitudinal (longitudinal stability 縦安定)
経由で〔前〕	via (telecast via satellite 衛星中継のテレビ放送)
経路 針路 コース〔名〕	path (approach path 進入経路)

結果 成果〔名〕 結果として生じる〔動〕	result (good results 良い結果) result
決定 決断 決意〔名〕 決定する 決心する〔動〕	decision (The pilot made a decision to land. パイロットは、着陸を決心した。) determine, decide
結合 組合せ〔名〕 結合する〔動〕	combination, linkage combine, link
結論〔名〕 結論を下す〔動〕	conclusion conclude
権威 権限 職権〔名〕	authority (execution of authority 職権の行使)
検査 点検 審査〔名〕 検査する 点検する 審査する〔動〕	inspection (bore-scope inspection ボアスコープによる検査) examination inspect, examine
検出〔名〕 検出する〔動〕	detection detect (detect a foreign substance 異物を検出する)
原因 根拠〔名〕 原因となる〔動〕	cause cause (Strong headwind caused arrival delay. 強い向かい風が原因で、到着が遅れた。)
現行の〔形〕	existent (existent circumstances 現行の情勢)
言語 言葉〔名〕	language (language habit 言語習慣)
現在の〔形〕	present (present situation 現状)
減少 減退 削減〔名〕 減少する〔動〕	abatement, reduction (cost reduction 経費削減) decrease abate, reduce, decrease
原則 原理 基本〔名〕	principle, fundamental

限度 限界〔名〕	limit, limitation (operating limitation 運用限界)
こ	
広域航法〔名〕	area navigation (RNAV)
降下〔名〕	descent (descent clearance 降下のクリアランス)
降下する〔動〕	descend (rapidly descend 急速に降下する)
交換 代替品〔名〕	replacement (交換 代替品) exchange
交換する〔動〕	replace (The maintenance person replaced the control unit 整備者が、制御器を交換した。) exchange (The pilots exchanged their seats. パイロットは、座席を交代した。)
航空 飛行〔名〕	aviation (civil aviation 民間航空)
航空の〔形〕	air (air transportation 航空輸送) aerial, aeronautical
航空電子工学の〔形〕	avionic
航空会社〔名〕	airline
航空機〔名〕	aircraft, airplane
航空機運航者〔名〕	aircraft operating agency
航空機から航空機へ	air to air (air to air communication パイロット相互間の通信)
航空機局〔名〕	aircraft station
航空機衝突防止装置〔名〕	traffic alert and collision avoidance system (TCAS)
航空機地球局〔名〕	aircraft earth station
航空局〔名〕	aeronautical station
航空局〔名〕	civil aviation bereau (CAB)
航空公衆通信〔名〕	aeronautical public communication (APC)
航空交通〔名〕	air traffic
航空交通管制〔名〕	air traffic control (ATC)
航空交通管理〔名〕	air traffic management (ATM)
航空交通管制センター〔名〕	area control center (ACC)

航空交通業務〔名〕	air traffic service (ATS)
航空固定業務〔名〕	aeronautical fixed service (AFS)
航空情報マニュアル〔名〕	aeronautical information manual (AIM)
航空地球局〔名〕	aeronautical earth station
航空保安施設〔名〕	air navigation aid
航空法〔名〕	civil aviation law (CAL)
航空法施行規則〔名〕	civil aviation regulation (CAR)
航空路〔名〕	airway, air route
航空路監視レーダー〔名〕	air route surveillance radar (ARSR)
航空路誌〔名〕	aeronautical information publication (AIP)
貢献 寄与〔名〕	contribution
貢献する 寄与する〔動〕	contribute (Tailwind contributed an early arrival. 追風が、早い到着に寄与した。)
行使 実行〔名〕	execution (execution of authority 権限の行使) performance (実行)
行使する 実行する〔動〕	execute, perform (perform duty 職務を実行する)
講習課程〔名〕	training course
高周波 短波〔名〕	high frequency (HF)
交信 通信〔名〕	contact, communication
交信する 通信する〔動〕	contact, communicate
構成 合成〔名〕	composition (crew composition 乗員編成)
構成する 合成する〔動〕	compose comprise (CNS comprises communication, navigation and surveillance. CNSは、通信、航法及び監視で構成される。)
合成の〔形〕	composite (composite flight HUD利用の飛行)

交通 通信〔名〕	traffic (heavy traffic 激しい交通) traffic (通信)
肯定的な〔形〕	affirmative (affirmative response 肯定的な応答)
交点 交線〔名〕	intersection (Intersection of distance lines makes a fix. 距離線の交点は、フィックスとなる。)
交差する 横切る〔動〕	intersect, cross, crosscut
口頭の 口述の〔形〕	verbal (verbal report 口頭の報告) spoken oral (oral exam 口述試験)
高度 標高〔名〕	altitude, elevation, height, level (sea level 海面)
飛行高度〔名〕	flight level (FL 370 飛行高度37,000フィート)
高度計〔名〕	altimeter (barometric altimeter 気圧高度計)
高度計規正(値)〔名〕	altimeter setting (Altimeter setting is obtainable through ATIS. 高度計規正値は、ATISから入手できる。)
行動 活動 動作 態度〔名〕	action (corrective action 修正動作) behavior
行動する〔動〕	act, behave
広範囲の〔形〕	wide (wide area, wide range 広範囲)
公表 発表〔名〕	publication, announcement, declaration
公表する 発表する〔動〕	publish, announce, declare, publicize (En-route frequencies are publicized. 航路上の周波数は、公表されている。)
公平〔名〕	fairness, equity
公平な 公正な〔形〕	fair (fair deal 公平な取扱い) equitable
公平に 公正に〔副〕	fairly, equitably

航法〔名〕	navigation
航行する〔動〕	navigate (navigate over the Pacific 太平洋上を航行する)
地文航法〔名〕	ground reference navigation
推測航法 (DR航法)〔名〕	dead reckoning navigation
天測航法〔名〕	astronomical navigation celestial navigation
無線航法〔名〕	radio navigation
自蔵航法〔名〕	self contained navigation
項目 条項〔名〕	item
項目別にする〔動〕	itemize
効率〔名〕	efficiency
効率的な〔形〕	efficient (efficient operation 効率的な運用)
効率的に〔副〕	efficiently
交流〔名〕	alternate current (AC)
航路〔名〕	course, route
航路上で〔副〕	en-route (en-route frequency 航路上での周波数)
超えて〔前〕	beyond (beyond the sea 海のかなた)
越える〔動〕	exceed (exceed the operating limit 運用限界を超える)
国際移動通信衛星機構〔名〕	International Mobile Satellite Organization (IMSO)
国際空港〔名〕	international airport
国際航空運送協会〔名〕	International Air Transportation Association (IATA)
国際的な〔形〕	international
国際的に〔副〕	internationally (internationally standardized terminology 国際的に標準化された用語)
国際電気通信連合〔名〕	International Telecommunication Union (ITU)

国際標準大気〔名〕	international standard atmosphere (ISA)
国際民間航空機関〔名〕	International Civil Aviation Organization (ICAO)
国籍〔名〕	nationality
国内の 自国の〔形〕	domestic (domestic operation 国内の運航) native (native language 自国の言葉)
個々 個人〔名〕 個々の〔形〕	individual individual (individual cases 個々の事例)
心に留めておく〔動〕	bear in mind
試み〔名〕 試みる〔動〕	attempt, trial attempt (The pilot attempted another contact. パイロットは、再度連絡を試みた。) try
答え 応答〔名〕 答える 応答する〔動〕	answer, response, reply answer, respond, reply
固定する 留める〔動〕 位置〔名〕 固定業務	fix, secure, fasten fix fixed service
異なる〔形〕	different (different frequency 異なる周波数)
言葉の 口頭の〔形〕 言葉で〔副〕	verbal (verbal report 口述報告) verbally
好ましい〔形〕 好む〔動〕	preferable (むしろ～の方が好ましい) prefer (むしろ～を好む) (Airline pilots prefer ILS to GCA. 航空会社のパイロットは、GCAよりもILSを好む。)
困難〔名〕 困難な 難しい〔形〕	difficulty difficult (difficult task 難しい仕事) hard (hard work 困難な仕事)

混雑〔名〕	congestion (traffic congestion 交通渋滞)
混雑した〔形〕	crowded (crowded terminal building 混雑したターミナルビルディング)
混雑する 混雑させる〔動〕	congest (Traffic congested the street. 街路は、交通で渋滞した。)
混乱〔名〕 混乱させる〔動〕	confusion, disorder confuse
語 用語 述語 専門語〔名〕	word, term
合理的な 論理的な〔形〕 合理的に 論理的に〔副〕	rational, reasonable, logical rationally, reasonably, logically
極超短波〔名〕	ultra high frequency (UHF)

さ

最後の 先程の 後者の〔形〕	last, final, latter
最終の〔形〕 最終進入〔名〕 最終進入経路〔名〕	final final approach final approach course
最初の〔形〕 最初に〔副〕	first, initial, original, earliest initially, originally, first, firstly
最新の〔形〕 最新化する〔動〕	latest (the latest version 最新版) up-date (The contents of the manual are up-dated. マニュアルの内容は、最新化されている。)
再送信する〔動〕	retransmit
再調査〔名〕 再調査する〔動〕	review, re-investigation review (ACC reviews the flight plan before issuing clearance. ACCは、クリアランス発行の前に飛行計画を再調査する。) re-investigate
最低気象条件〔名〕	weather minima (weather minima for takeoff 離陸気象要件)

差異 相違〔名〕 　異なる 相違する〔形〕 　違って〔副〕	difference different differently
裁量 判断〔名〕	discretion (pilot's discretion パイロットの裁量)
先の〔形〕	further (further study 今後の研究)
先んじる 優先する〔動〕 　先行する〔形〕	precede (Flight safety precedes on-time operation. 安全飛行は、定刻運航に先んじる。) preceding (preceding aircraft 先行する航空機)
削減 軽減 カットバック〔名〕	cut-back (power cut-back for noise abatement 騒音軽減のためのパワーカットバック) reduction in workload (作業量の軽減)
避ける 回避する〔動〕	avoid (avoid cloud 雲を回避する)
誘い 招待〔名〕 誘う 招待する〔動〕	invitation (Go-ahead is a word of invitation to communication. Go-ahead は、通信への誘いの語である。) invite
作動 活動的にすること〔名〕 作動する 作動させる〔動〕	operation, activation operate (Engines are operating normally. エンジンは、正常に作動している。) activate (Fire warning was activated. 火災警報が、作動した。)
さまざまな〔形〕	various (Pilots suggested various opinions on fuel saving. パイロットは、燃料節減に関するさまざまな意見を提案した。)
参加〔名〕 参加する〔動〕	participation participate (participate in a campaign 運動に参加する) join (join the activities 活動に参加する)

参考 参照 言及〔名〕	reference (reference material 参考資料)
参考にする 言及する〔動〕	refer (refer to the manual マニュアルを参考にする)
算定 計算〔名〕	calculation, count
算定する 計算する〔動〕	calculate (DME calculates slant distance to the station. DMEは、局までの斜距離を算出する。) compute, count
残存する〔動〕	remain
残り〔名〕	remainder (remainder of the food 食べ物の残り) rest (The rest of the pages explain weather phenomena. 残りの頁は、気象現象を説明する。)
残りの〔形〕	remaining (remaining fuel 残存燃料)

し

支援 支持〔名〕	support, help, aid, assistance
支援する 支持する〔動〕	support, help, aid, assist
視界 視野 視力 視覚〔名〕	view, eyesight, sight (The runway is in sight, 滑走路は、見えている。) vision (field of vision 視野)
資格〔名〕	qualification
資格を与える〔動〕	qualify
指揮 監督 管制 先導〔名〕	command, control, lead, direct
指揮する〔動〕	direct, control, command
試験〔名〕	test, examination
試験飛行〔名〕	test flight
試験する〔動〕	examine, test
資源〔名〕	resource (natural resource 天然資源)
指示 命令〔名〕	instruction, direction
指示する 指導する〔動〕	instruct, direct
指示対気速度〔名〕	indicated airspeed (IAS)

沈む〔動〕	sink (sink rate 沈下率)
施設 設備〔名〕	facility (radio navigation facility 無線航法施設)
～に従う〔動〕	follow (follow the instruction 指示に従う) obey comply (comply with the clearance クリアランスに従う)
しっかりした 安定した〔形〕	firm (firm touch down しっかりした着陸) steady (steady progress 安定した進歩)
しっかりと〔副〕	firmly, steadily
失敗〔名〕	failure
失敗する 失念する〔動〕	fail, lack (lack of attentiveness 注意の欠落)
質問〔名〕	question
質問する〔動〕	question, interrogate (interrogator 質問機)
質問事項 アンケート〔名〕	questionnaire
視程 視界〔名〕	visibility
見える〔形〕	visible (The runway was visible from far distance. 滑走路は、遠くから見えていた。)
示す〔動〕	indicate, show
斜距離〔名〕	slant distance
尺度 縮尺 スケール〔名〕	scale
車両 乗り物〔名〕	vehicle (The spacecraft is a vehicle. 宇宙船は、乗り物である。)
周期的な 定期的な〔形〕	periodic
周期的に 定期的に〔副〕	periodically (ATIS is periodically updated. ATISは、定期的にアップデートされる。)
終結〔名〕	termination, end

終結する 終わる〔動〕 終わりの〔形〕	end, terminate end (end result 最終結果)
集団 群れ〔名〕	group (A group of birds are flying near the runway. 鳥の群れが、滑走路の近くを飛んでいる。)
集中 集中力 濃縮〔名〕 集中する 濃縮する 専心する〔動〕	concentration concentrate (Pilots concentrate themselves on maneuvering the aircraft during takeoff. パイロットは、離陸中は操縦に集中する。) devote, focus
周波数〔名〕	frequency (frequency band 周波数帯)
終了 終り〔名〕 終了する〔動〕	termination (termination of message 通報の終了) end, finish terminate, end, finish out (通信の終了を意味するICAO設定の標準用語)
種々の〔形〕	multi (multi-purpose satellite 多目的衛星)
受信証〔名〕	acknowledgement of receipt
手段〔名〕	means (a means of communication 通信手段)
出発〔名〕 出発する〔動〕 出発前の〔形〕 出発予定時間〔名〕	departure (departure frequency 出発時の周波数) depart, leave pre-departure (pre-departure briefing 出発前のブリーフィング) estimated time of departure (ETD)
出域管制 出域管制周波数	departure control departure frequency

し

取得 獲得〔名〕 取得する 獲得する〔動〕	acquisition acquire, gain, obtain
主要な〔形〕	main (main part 主要部分) major (major airlines 主要航空会社) leading, principal
障害 障害物〔名〕 妨害する〔動〕	obstruction obstruct
衝撃波〔名〕	shockwave (A shockwave is generated at the speed of sound. 音速になると衝撃波が生じる。)
小数点〔名〕	decimal
衝突〔名〕 衝突する〔動〕	collision (ATC protects aircraft from collision. ATCは、航空機を衝突から守る。) collide
承認 了承 領収〔名〕 承認する 了承する 領収する〔動〕 受信証〔名〕	acknowledgement, approval acknowledge (The pilot acknowledged the revised clearance. パイロットは、改定クリアランスを了承した。) approve acknowledgement of receipt
初心者 未熟者〔名〕	amateur (アマチュア) beginner (初心者) novice (未熟者)
書類 文書〔名〕	document
知らせ 通知 告示〔名〕 知らせる 通知する 告示する〔動〕	information, notification (通知 告示) inform, notify (告示する)
調べる 検査する〔動〕	check, inspect, examine
資料 データ〔名〕	data (flight data recorder 飛行データ記録装置) material (資料)

し

知って 気づいて〔形〕	aware (Pilots are well aware of the weather along the flight route. パイロットは、飛行コース上の気象について十分知っている。)
指令 命令〔名〕	order, command
指令する 指図する〔動〕	order, command
進歩 発達〔名〕	progress, development
進歩する 発達する〔動〕	progress, advance, develop (developing country 発展途上国)
信号〔名〕	signal
信号波〔名〕	signal wave (A signal wave modulates the carrier wave. 信号波が、搬送波を変調する。)
真対気速度〔名〕	true airspeed (TAS) (The speed used in the flight plan is the true air speed. 飛行計画で使用される速度は、真対気速度である。)
進入〔名〕	approach
進入管制〔名〕	approach control
進入経路〔名〕	approach course
進入復行〔名〕	missed approach
進入方式〔名〕	type of approach
振幅〔名〕	amplitude
振幅変調〔名〕	amplitude modulation (AM)
進歩 発達〔名〕	advance, advancement, progress
進歩的な 斬新的な〔形〕	progressive (progressive change 斬新的変化)
信頼 信頼性 確実性〔名〕	reliance, reliability (信頼性) trust (信頼)
信頼できる〔形〕	reliable (reliable equipment 信頼性の高い機器)
信頼する〔動〕	rely, trust
磁気 磁性 磁気作用〔名〕	magnetism
磁気の 磁気を帯びた〔形〕	magnetic (magnetic field 磁場)

磁気嵐〔名〕	magnetic storm
事前に 前もって〔副〕	prior to (The pilot informed the maintenance section of a generator failure prior to his arrival. パイロットは、発電機の故障を到着前に整備に知らせた。 beforehand (The pilot reviewed the descent profile beforehand. パイロットは、前もって降下中の状況を調べた。) in advance
前の 以前の 先行する〔形〕	prior, previous, preceding (preceding aircraft 先行機)
持続 持続能力 耐久性〔名〕	endurance (耐久性) continuation (持続)
滞空飛行〔名〕	endurance flight
事態 状態〔名〕	situation, status, condition (favorable weather conditions 順調な気象状態)
実行 遂行〔名〕	execution
実行する 遂行する〔動〕	execute (execute go-around 着陸復行を実行する) conduct, perform
自動的な 機械的な〔形〕	automatic
自動的に 機械的に〔副〕	automatically
自動従属監視〔名〕	automatic dependent surveillance (ADS)
自動制御（自動制御機器）〔名〕	automatic control (automatic controller)
自動操縦装置〔名〕	autopilot
自動方向探知機〔名〕	automatic direction finder (ADF)
従事する 関係する 参加する〔動〕	engage (engage in flight operation 運航に従事する) involve, participate
充足 満足〔名〕	satisfaction, sufficiency
満足させる〔動〕	satisfy

渋滞〔名〕	congestion (traffic congestion 交通渋滞)
渋滞させる〔動〕	congest (Traffic congested the street. 道路は、交通で渋滞した。)
重大な 厳しい〔形〕	serious (serious damage 重大な損害) severe, grave (grave information 重大な情報)
重要性〔名〕	importance, significance
重要な 著しい〔形〕	important (important matter 重要な事柄) significant (a significant change in headwind component 向かい風成分の著しい変化)
従来の〔形〕	conventional (conventional radio equipment 従来型の無線機器)
受信〔名〕	reception
受信する〔動〕	receive (receiver 受信機)
受信局〔名〕	receiving station
準備〔名〕	preparation
準備する〔動〕	prepare (The dispatcher has prepared the flight plan. ディスパッチャーは、飛行計画を準備した。)
乗客〔名〕	passenger (passenger seat 乗客の席)
状態 環境〔名〕	circumstance, situation condition (flight condition 飛行状態)
条件 要件〔名〕	condition (impose conditions 条件をつける) requirement (要件)
条件として〔形〕	subject to (subject to the controller's approval 管制官の承認を条件として)
上限〔名〕	upper limit
場周経路〔名〕	traffic pattern (left traffic pattern 左旋回の場周経路)

上昇〔名〕 上昇する〔動〕	climb climb (climb to the assigned altitude 指定高度へ上昇する)
上手 熟練〔名〕 上手な 熟練した〔形〕	skill, proficiency skilled, skillful, proficient
状態 状況〔名〕	condition, situation
冗長 余剰〔名〕 冗長な〔形〕	redundancy redundant (redundant installation 多重装備)
情報〔名〕	information (weather information 気象情報)
乗務員〔名〕	flight crew (a flight crew member 個々の乗務員)
条例 法令 規定〔名〕	ordinance, regulation
除去 削除 排除〔名〕 除去する 削除する 排除する〔動〕	elimination, deletion, exclusion (排除) eliminate, delete, exclude (排除する)
助言〔名〕 助言する〔動〕	advice (The captain followed the co-pilot's advice. 機長は、副操縦士の助言に従った。) advise
序文 まえがき〔名〕	preface, introduction
迅速な 機敏な〔形〕 迅速に 機敏に〔副〕	prompt (prompt action 機敏な行動) rapid (迅速な) promptly, rapidly

す

衰弱 減衰〔名〕 減衰する 弱まる 弱める〔動〕	attenuation attenuate (Radio waves attenuate during propagation. 電波は、伝播中に減衰する。) weaken (弱まる 弱める)

推定〔名〕 推定する〔動〕	estimate, presumption estimate (The pilot roughly estimated the remaining fuel over the next fix. パイロットは、次の地点における残存燃料を概算した。) presume (推定する)
推力〔名〕	thrust (thrust of a jet engine ジェットエンジンの推力)
数字〔名〕 数の 数値方式の〔形〕	digit, figure (double figures 2桁の数字 three figures 3桁の数字) numeral, numerical
少なくとも とにかく〔副〕	least (at least 少なくとも、とにかく)
少し わずかの〔形〕 少し わずかに〔副〕	slight, little slightly (The path is slightly higher than the glide slope. パスが、グライドスロープよりわずかに高い。)
進む 進める〔動〕	proceed (proceed to the next phase 次の段階へと進む) advance (advance the thrust levers スラストレバーを進める)
すばやい〔形〕 すばやく〔副〕	rapid (rapid glance すばやい一瞥) quick rapidly (rapidly descendすばやく降下する) quickly
スピードブレーキ〔名〕	speed brake (The speed brakes spoil the lift of the wing. スピードブレーキは、翼の揚力を損失させる。)
～すべき〔助〕	shall (法律、規則の条文で、～すべきである) (The flight clearance shall be read back by the pilot. フライトクリアランスは、パイロットによって復唱されるべきである。)

す

すべての〔形〕	every (どの〜もみな) all (すべての) whole (全体の) entire (全部の)
スペクトル 範囲〔名〕	spectrum
スラストレバー〔名〕	thrust lever (推力制御レバー)

せ

正確 正確さ 精密 精密さ〔名〕	accuracy, precision
正確な 精密な〔形〕	accurate, precise, exact
正確に 精密に〔副〕	accurately, precisely, exactly
制限 限定 限界〔名〕	restriction, limit, limitation
制限する 限定する〔動〕	restrict, limit
静止衛星〔名〕	geo-stationary satellite
晴天〔名〕	clear sky
晴天乱気流〔名〕	clear air turbulence (CAT)
政府 政府機関〔名〕	government (government agency政府機関)
性 性別〔名〕	sex, gender
精密〔名〕	precision
精密な〔形〕	precise, precision, exact
精密に〔副〕	precisely, exactly
精密進入経路指示灯〔名〕	precision approach path indicator (PAPI)
精密進入着陸〔名〕	precision approach and landing
精密進入レーダー〔名〕	precision approach radar (PAR)
整流器〔名〕	rectifier
整流する〔動〕	rectify (AC is rectified to DC. 交流は、直流に整流される。)
整列する〔動〕	line up (The tower instructed the pilot to line up and hold on the runway. タワーは、パイロットに滑走路上で静止するように指示した。)

せ

世界〔名〕	world (worldwide network世界的なネットワーク)
世界協定時〔名〕	coordinated universal time (UTC)
積雲〔名〕	cumulus
積乱雲〔名〕	cumulonimbus (Cb)
赤道〔名〕	equator (MTSAT are located over the equator. MTSATは、赤道上空に位置している。)
責任 信頼性 反応〔名〕	responsibility, response (a quick response by the receiving station 受信局によるすばやい反応)
責任のある〔形〕	responsible (Pilots are responsible for safe operation of flight. パイロットは、飛行の安全な実施に責任がある。)
積極的な 活動的な〔形〕	active (active volcano 活火山)
積極的に 活発に〔副〕	actively
接近〔名〕	approach
接近する 出入りする 近づく〔動〕	approach (approach the runway 滑走路に近づく)
	access (Access to the cockpit in flight is prohibited. 飛行中操縦室への出入りは、禁止されている。
接近度〔名〕	closure rate
接地〔名〕	touchdown (firm touchdown しっかりした接地)
接地する〔動〕	touchdown
接地点〔名〕	touchdown point
接地帯〔名〕	touchdown zone
設定 設立 施設〔名〕	establishment (communication establishment 通信設定)
設立する 設定する〔動〕	establish
切迫した 緊急の 差し迫った〔形〕	imminent, urgent, impending (impending hot start 差し迫ったホットスタート)

接続〔名〕 接続する 連結する〔動〕	connection connect, link
セルコール 選択呼出装置〔名〕	selective calling system (SELCAL)
線 行 路線〔名〕	line (domestic lines 国内線)
旋回〔名〕 旋回する〔動〕	turn turn (turn right 右に旋回する)
宣言 布告 公表 発表〔名〕 宣言する 公表する〔動〕	declaration, publication, announcement declare, announce, publish
選択 取捨〔名〕 選択する 選ぶ〔動〕	selection, option, choice select, choose
絶対的な〔形〕 絶対的に〔副〕	absolute (absolute priority 絶対的優先権) absolutely (Smoking is absolutely prohibited. 喫煙は、絶対禁止されている。)
全体の〔形〕	entire (It snowed the entire day. 一日中雪が降った。) whole (the whole world 全世界)
全般の 包括的な 全体的な〔形〕 全世界の〔形〕	general, generic, global universal, global (global positioning system 全世界測位システム)
前方へ〔副〕	ahead (go ahead 通信継続を促す用語)

そ

相違〔名〕	difference (difference in language habit 言語習慣の相違)
相違する〔形〕	different (The pilot contacted a different station. パイロットは、異なった局に連絡した。)
相違する〔動〕	differ

騒音 音〔名〕	noise
騒音軽減方式〔名〕	noise abatement procedure
総括呼出〔名〕	general call
遭遇 出会い〔名〕	encounter
遭遇する 出会う〔動〕	meet, encounter (The pilot encountered CAT. パイロットは、CATに遭遇した。)
走行区域〔名〕	maneuvering area (Traffic on the maneuvering area is monitored by ASDE. 走行区域の交通は、ASDEによって監視されている。)
操作 取扱い〔名〕	operation, handling, manipulation
操作する 取り扱う 処理する〔動〕	operate, handle, manipulate, manage
捜索 追及〔名〕	search
捜索する〔動〕	search (search for the missing aircraft 行方不明の航空機を捜索する)
捜索救難〔名〕	search and rescue (SAR)
総称的な〔形〕	generic, general, common
送信〔名〕	transmission (Transmission power is enhanced. 送信電力は、高められた。)
送信する〔動〕	transmit
送信系統〔名〕	transmitting system
送信方法 送信手順〔名〕	transmitting procedure
操縦〔名〕	maneuver, control
操縦する〔動〕	maneuver, control
操縦桿〔名〕	control column
操縦室 コックピット〔名〕	cockpit, flight deck
総体の 全体の〔形〕	gross, whole (全体の)
総重量〔名〕	gross weight (gross takeoff weight 離陸総重量)
装置、装備〔名〕	apparatus, device, equipment, installation

遭難 惨事〔名〕	distress, disaster
遭難状態〔名〕	distress condition
遭難通信〔名〕	distress traffic
遭難通報〔名〕	distress message
遭難呼出〔名〕	distress call
即座の 早速の〔形〕	immediate (immediate answer 即答)
即座に 早速〔副〕	immediately, promptly (The pilot promptly raised the pitch. パイロットは、直ちにピッチを上げた。)
促進する 寄与する〔動〕	expedite, promote (The stabilized approach concept promotes flight safety. 安定したアプローチの考えは、飛行の安全に寄与する。)
測定 寸法〔名〕	measurement, measure
測定する〔動〕	measure (The low range radio altimeter measures the height of an aircraft below 2,500 feet. 低高度電波高度計は、2,500フィート以下で機の高度を測定する。)
速度〔名〕	speed, velocity (wind velocity 風速)
損なう〔動〕	miss (The pilot missed the speed nearing the limit. パイロットは、速度が限界に近づきつつあることを見落とした。) lose (The pilot lost vision by the glare. パイロットは、まぶしさで先が見えなくなった。)
組織 構成〔名〕	organization
組織する 編成する〔動〕	organize
～に沿って〔前〕	along (flight conditions along the course コース上の飛行状態)
そばで〔前〕	beside (CAT often exists beside the jet stream. CATは、しばしばジェットストリームのそばに現れる。)

た

存在〔名〕	existence (Existence of lightning was reported by a pilot. 雷光の存在が、パイロットにより報告された。)
存在する 現れる〔動〕	exist
増加〔名〕	increase
増加する〔動〕	increase (Misunderstandings of communication between controllers and pilots have increased. 管制官とパイロットの間の通信の誤解が、増えている。)

た

タービン〔名〕	turbine (turbine engine タービンエンジン)
ターミナル 終点 起点〔名〕	terminal
ターミナル管制区〔名〕	terminal control area (TCA)
待機 ホールディング〔名〕	holding (holding procedure 待機の方式)
スタンドバイ〔名〕	standby
待機する〔動〕	hold (hold over the next fix 次の地点上空で待機する) standby
対抗機位置情報〔名〕	traffic advisory (TA)
対地接近警報システム〔名〕	ground proximity warning system (GPWS)
対地速度〔名〕	ground speed (GS)
大洋 海洋〔名〕	ocean (fly over the ocean 洋上飛行)
大洋の 大洋性の〔形〕	oceanic (oceanic climate 海洋性気候)
対流圏〔名〕	troposphere
高い〔形〕	high (high frequency 短波)
タクシー〔名〕	taxi (taxi way 誘導路)
タクシーする〔動〕	taxi (taxi on the maneuvering area 走行区域をタクシーする)

タクシークリアランス〔名〕	taxi clearance
確かめる〔動〕	confirm (confirm the clearance by read-back 復唱してクリアランスを確かめる) verify
確認〔名〕	confirmation, verification
正しい〔形〕	correct (correct decision 正しい決定) right (right answer 正しい回答)
直ちに〔副〕	immediately (The pilot immediately advanced the thrust levers. パイロットは、直ちにスラストレバーを進めた。)
即時の〔形〕	immediate (immediate response 即座の応答)
縦の〔形〕	vertical (The vertical distance between the surface and the cloud base is the ceiling. 地表面と雲底の間の縦の距離は、雲高である。)
縦に〔副〕	vertically
短距離〔名〕	short range (short range aircraft 短距離用航空機) short distance
探知〔名〕	detection (detection by radar レーダーによる探知)
探知する〔動〕	detect (detect hazardous aircraft 危険な航空機を探知する)
探知可能な〔形〕	detectable (detectable range 探知可能範囲)
探知装置〔名〕	detective device
短波〔名〕	short wave, high frequency (HF)
代表的な 特有の〔形〕	typical (typical weather condition 特有の気象状態)
妥協〔名〕	compromise
妥協する〔動〕	compromise (Radar detectability compromises with frequencies to be used. レーダーの探知能力は、使用周波数との妥協である。)

た

～だけれども〔接〕	although (The pilot took off although the destination weather was uncertain. 目的地の気象ははっきりしなかったけれども、パイロットは離陸した。)
誰でも〔代〕	anyone (Anyone may participate in this training course. 誰でもこの訓練に参加できる。)
段階 位相〔名〕	phase (flight phase 飛行中の一つの段階) variable phase (可変位相)

ち

遅延 遅滞〔名〕 　遅らせる〔動〕	delay (without delay 即刻) delay (The flight 101 is delayed. 101便は、遅れている。)
近い〔形〕 近く〔前〕 近くの〔形〕 近くで〔副〕 近づく 接近する〔動〕	near (near to the runway 滑走路に近い) (near the runway 滑走路の近くに) nearby (nearby airport 近くの空港) nearby (The airport is nearby. 空港は、近くである。) near (A front is nearing. 前線が近づきつつある。) approach (The pilot approached the runway through turbulent air. パイロットは、乱気流の中を通って滑走路に近づいた。)
地形 地勢〔名〕	terrain (The airborne weather radar displays terrain clearly. 機上の気象レーダーは、地形をはっきりと表示する。)
地上員〔名〕	ground personnel, ground crew
地上局〔名〕	ground station
地上誘導着陸方式〔名〕	ground controlled approach (GCA)

地帯 区域〔名〕	zone (touchdown zone 接地帯)
秩序〔名〕	order, regularity
秩序ある 整然とした〔形〕	orderly (orderly flow of traffic 交通の秩序ある流れ)
地方の〔形〕	local (local weather 地方の気象状態) regional (regional accent 地方の方言)
地方時〔名〕	local time
着氷〔名〕	icing (icing condition 着氷状態)
着陸〔名〕	landing (smooth landing スムーズな着陸)
着陸する〔動〕	land (land on time 所定時刻に着陸する)
着陸滑走〔名〕	landing roll
着陸装置〔名〕	landing gear
着陸復行〔名〕	go-around
チャンネル 割当周波数帯〔名〕	channel
注意〔名〕	attention, attentiveness (lack of attentiveness 注意不足) caution (with caution 用心して)
注意深い〔形〕	careful, attentive
注意深く〔副〕	carefully, attentively
不注意な〔形〕	careless, inattentive
不注意に うっかり〔副〕	carelessly, inattentively
中間点〔名〕	midpoint
駐機する〔動〕	park (park the aircraft 航空機を駐機する)
駐機場〔名〕	parking spot, parking area
中継〔名〕	relay
中継する 取り次ぐ〔動〕	relay (relay the message 通報を中継する)
中止 解消〔名〕	discontinuance, suspension, cancellation
中止する 解消する〔動〕	abort (abort takeoff 離陸を中断する) discontinue, suspend, cancel

ち

中心線〔名〕	centerline (runway centerline 滑走路の中心線)
中程度の 穏やかな〔形〕	moderate (moderate turbulence 中程度の乱気流) mild (mild climate 穏やかな気候) gentle medium (medium size 中型)
中波〔名〕	medium frequency (MF)
超過する〔動〕	exceed (exceed the speed limit 制限速度を超過する)
長距離〔名〕	long distance (long distance flight 長距離飛行)
調査〔名〕 調査する〔動〕	investigation (accident investigation 事故調査) examination (under examination 調査中) investigate, examine
調節 調整〔名〕 調節する〔動〕	adjustment adjust (adjust the frequency 周波数を合わせる)
超短波〔名〕 超短波全方向式無線標識〔名〕	very high frequency (VHF) VHF omni-directional radio range (VOR)
超長波〔名〕	very low frequency (VLF)
ちょうど〔形〕	whole (a whole hundred ちょうど百)
長波〔名〕	low frequency (LF)
長文通報〔名〕	long message
眺望 視界 見解〔名〕	view, scene, sight, opinion (take a general view of ～を概観する)
調和 比率〔名〕 調和して〔形〕 調和する 匹敵する〔動〕	proportion, consistency consistent match, suit, fit, meet

直接の〔形〕	direct (direct call 直接呼出し)
直接に〔副〕	directly (fly directly 直行する)
直前に〔副〕	just before, immediately before, directly before
直流〔名〕	direct current (DC) (Radio apparatus operates on DC. 無線装置は、直流で作動する。)
沈黙〔名〕	silence (The word MAYDAY imposes communication silence to all stations. MAYDAYの語は、すべての局に対して通信の沈黙を課する。)
無言の〔形〕	silent (silent reading 黙読)
無言で〔副〕	silently

つ

追加〔名〕	addition (There was a gust of wind in addition to heavy rain. ひどい雨に加えて突風もあった。)
追加の〔形〕	additional (The pilot requested additional fuel. パイロットは、追加の燃料を要求した。)
追加的に〔副〕	additionally
費やす〔動〕	spend, consume (consumer 消費者)
消費〔名〕	consumption (fuel consumption rate 燃料消費率)
通過〔名〕	pass, transit (transit passenger 通過客)
通過する〔動〕	pass
通信 文通〔名〕	communication, correspondence
通信する　文通する〔動〕	communicate, correspond
通信設定〔名〕	communication establishment
通信途絶〔名〕	communication failure
通信不能〔名〕	lost communication
通信機器〔名〕	communication equipment (apparatus)
通信手順〔名〕	communication procedure

通常 正常 標準〔名〕	normal (return to normal operation 通常操作に戻る)
通常の 標準の〔形〕	normal, usual
通常操作〔名〕	normal procedure
故障時操作〔名〕	abnormal procedure
緊急時操作〔名〕	emergency procedure
通常は 一般に〔副〕	normally, usually, commonly
通報〔名〕	message, information, notice (NOTAM is a notice to airmen. NOTAMは、運航関係者に対する通報である。)
使う 利用する 適用する〔動〕	use, utilize, apply
使用〔名〕	use (use of the standard phraseology 標準用語の使用) application (application of advanced technique 最新技術の適用)
次の〔形〕	next, following (the following aircraft 次に続く航空機)
作り出す〔動〕	make, produce (The oscillator produces a carrier wave. 発振器は、搬送波を作り出す。)
綴る〔動〕	spell (spell a word 語を綴る) write (write a sentence 文を綴る)
強さ〔名〕	strength (strength of a light 光の強さ) intensity (the degree of intensity 強さの度合い)
強い〔形〕	strong, powerful (powerful engine 力強いエンジン)
強く〔副〕	strongly, powerfully (Jet engines powerfully drive heavy aircraft forward. ジェットエンジンは、重い航空機を強力に推進する。)
貫く〔動〕	penetrate (VHF radio waves penetrate the ionosphere. VHFの電波は、電離層を貫く。)

つ

て

提案〔名〕	proposal, suggestion
提案する 示唆する〔動〕	propose, suggest (The weather condition suggests holding or a diversion. 天気の状態は、待機かまたはダイバートすることを示唆している。)
定期的な 規則的な〔形〕	periodic, regular
定期的に 規則的に〔副〕	regularly, periodically (ATIS is periodically updated. ATISは、規則的にアップデートされる。)
提供 規定〔名〕	provision
提供する〔動〕	provide, supply, furnish
規定する 明記する〔動〕	stipulate (The operations manual stipulates the importance of flight safety. 運航規程は、安全飛行の重要性を明記している。) provide (The aircraft operating manual provides various abnormal procedures. 航空機運用規程は、種々の故障時操作を規定している。)
定期旅客機〔名〕	airliner (twin jet airliner 双発ジェット旅客機)
定義〔名〕	definition
定義する〔動〕	define
低高度〔名〕	low altitude, low level
低高度電波高度計〔名〕	low range radio altimeter (LRRA)
低高度突風〔名〕	low level windshear
停止〔名〕	stop, standstill (come to a stop 停止する)
停止する〔動〕	stop (stop line 停止線)
訂正 修正 改定〔名〕	correction, amendment
訂正する 修正する 改定する〔動〕	correct (correct the message 通報を訂正する) amend (amended route 改定後のルート)

低電圧〔名〕	low voltage (Low voltage is applied to radio apparatus. 無線機器には、低電圧が適用される。)
定例の〔形〕	regular (regular flight 定期便)
適切な〔形〕 適切に〔副〕	appropriate (appropriate corrective action 適切な修正動作) appropriately (appropriately manage the speed 速度を適切に処理する)
適法の〔形〕 非合法的な〔形〕	lawful (a lawful action 適法な行為) unlawful (unlawful entry 不法侵入) intrusion (侵入)
適用 出願 願書〔名〕 適用する 用いる〔動〕	application use, employ, apply (Pilots make efforts to apply stabilized approach concept to practical operation. パイロットは、安定したアプローチの考えを実際の運航に適用すべく努力する。)
手順 方式〔名〕	procedure (communication procedure 通信手順) (instrument approach procedure 計器進入方式)
手前の〔形〕	short (Taxi is limited to the stop line short of the runway. タクシーは、滑走路手前の停止線までに制限されている。)
点 小数点〔名〕	point (decimal point 小数点)
転移経路〔名〕	transition route (A transition route connects SID to an airway. 転移経路は、SIDと航空路を結ぶ。)
転換〔名〕 転換する〔動〕	conversion, changeover convert (The antenna converts electric current to a radio wave. アンテナは、電流を電波に転換する。)
データリンク〔名〕	data link

て

和	英
出来るだけ早く〔副〕	as soon as possible (begin descent as soon as possible 出来るだけ早く降下を開始する)
デジタル方式の〔形〕	digital (digital display デジタル表示)
デリンジャー現象〔名〕	Dellinger phenomenon
電圧〔名〕	voltage (high voltage 高電圧)
電気〔名〕 電気の〔形〕 放電 電流 電力	electricity electric (electric discharge 放電) (electric current 電流) (electric power 電力)
電気通信〔名〕	telecommunication (telecommunication network 電気通信網)
電源〔名〕	(electric) power source
電信〔名〕	telegraphy
電磁波〔名〕	electro-magnetic wave (A radio wave is an electro-magnetic wave. 電波は、電磁波である。)
伝達〔名〕 伝達する〔動〕	delivery, transmission transmit, deliver (deliver the clearance クリアランスを伝達する)
電波〔名〕 電波法〔名〕	radio wave (Radio waves propagate without vehicle. 電波は、媒体なしで伝播する。) radio law
伝播〔名〕 伝播する〔動〕	propagation (propagation of radio waves 電波の伝播) propagate (Radio waves propagate far distance. 電波は、遠くまで伝播する。)
電離層〔名〕	ionosphere (Some radio waves are reflected by the ionosphere. ある電波は、電離層で反射される。)

電力供給〔名〕	(electric) power supply
と	
当然の〔形〕	due (due consideration 十分の考慮)
到着〔名〕 到着する〔動〕 到着予定時刻〔名〕	arrival arrive estimated time of arrival (ETA)
登録 記録〔名〕 登録する 記録する〔動〕	registration, register, record register, record
遠い〔形〕 遠くへ〔副〕	far (far distance 遠距離) far ahead (The preceding aircraft has gone far ahead. 先行機は、ずっと遠くへ行った。)
時々〔副〕	sometimes, now and then, often, frequently occasionally (It will occasionally rain. 時々雨が降るでしょう。)
特性周波数〔名〕	characteristic frequency
特徴 特性〔名〕 特徴的な〔形〕	feature, character characteristic
特定の 特有の 特別の〔形〕	specific, particular, special
取って代わる〔動〕	supersede (The revised clearance supersedes the previously issued clearance. 改定クリアランスは、先に発行されたクリアランスに取って代わる。)
取扱い〔名〕 取り扱う〔動〕	handling, treatment handle, treat, deal with (deal with a difficult matter 難しいことを処理する)
取消し 廃止 失効〔名〕 取り消す〔動〕	cancellation, revocation, annulment cancel, revoke, annul
度 程度〔名〕	degree (OAT is −55℃. 外気温度は、マイナス55℃である。)

	degree (degree of signal strength 信号強度の程度)
同時間飛行可能点〔名〕	equal time point (ETP) (We've past the ETP. 同時間飛行可能点を通過した。)
同軸ケーブル〔名〕	coaxial cable (The coaxial cable is an excellent feeder. 同軸ケーブルは、優れた給電線である。)
同調〔名〕	synchronization synchronizer (同調器) tuning (波長の調整 (無線機 楽器))
同調する 同調させる〔動〕	synchronize (Radar scope sweeping is synchronized with the direction of the antenna. レーダースコープの走査は、アンテナの方向と同調している。)
同等の ～に相当する 類似の〔形〕	equivalent, similar, comparable (匹敵した)
同様の 同じ～〔形〕	same (the same altitude 同じ高度)
導入〔名〕	introduction
導入する〔動〕	introduce (introduce the new technology 新しい科学技術を導入する。)
独自の〔形〕	own (their own language habit 彼ら独自の言語習慣)
所有する〔動〕	own (The airline owns many aircraft. その航空会社は、多くの航空機を所有している。)
どちらか一方の どちらの～でも	either (The good weather allowed the pilot to choose either ILS or visual approach procedure. 天気が良くてパイロットは、ILSと目視のどちらの方式のアプローチでも選択することが出来た。)
努力〔名〕	effort (with effort 努力して) endeavor exertion (The task requires exertion. その仕事は、努力を要する。)

努力する〔動〕	make efforts, make exertion, endeavor
どれでも〔形〕	any (Any person can participate in this meeting. 誰でもこの会に参画できる。)
な	
～でない限り もし～でなければ〔接〕	unless (The controller will not issue takeoff clearance unless the weather minima is met. 管制官は、最低気象条件が満たされていない限り離陸許可は発行しない。)
内容 詳細〔名〕	content, detail (details of the analysis report 解析報告の詳細)
なおその上〔副〕	as well (He speaks English as well. 彼は、英語も話す。)
長い〔形〕	long (long message 長文の通報)
流れ〔名〕	flow (traffic flow 交通の流れ)
～なしに〔前〕	without (without effort 難なく)
なぜなら ～だから〔接〕	because (The flight was cancelled because it snowed. 雪が降ったのでフライトはキャンセルされた。)
なめらかな 平坦な〔形〕 なめらかに〔副〕	smooth (smooth runway 平坦な滑走路) smoothly (The aircraft landed smoothly 機は、スムーズに着陸した。)
～になる〔動〕	become (It is becoming colder. 寒くなりつつある。)
～から成る〔動〕	consist (The tower consists of three sections. タワーは、三つのセクションから成る。)
～になるまで〔前〕	until (Until then it had been snowing. その時までは、雪が降っていた。)

に

二次の〔形〕	secondary
二次レーダー〔名〕	secondary radar (The secondary radar sends a signal to a specific target. 二次レーダーは、特定の目標に対して信号を送る。)
二次監視レーダー〔名〕	secondary surveillance radar
偽の〔形〕	pseudo (pseudo satellite station 擬似衛星局)
日本標準時〔名〕	Japan standard time (JST)
入手する 捕捉する〔動〕	obtain (obtain permission 許可を得る) acquire (acquire a target 目標物を捕捉する) (acquire wind variation data 風の変化のデータを入手する)
任意の〔形〕	optional (fly an optional route by RNAV RNAVで任意のルートを飛行する)
認定 承認 許可〔名〕	approval, authorization
認定する 承認する 許可する〔動〕	approve (The controller approved the pilot's request of altitude change. 管制官は、パイロットの高度変更要求を承認した。) authorize

ね

～ねばならない〔助〕	must, have to, shall (法律、規則の条文で、～すべきである)
燃料〔名〕	fuel (fuel system 燃料系統)
燃料欠乏〔名〕	minimum fuel
燃料放棄〔名〕	fuel dumping, fuel jettisoning
年齢〔名〕	age (age deterioration 経年劣化)

の

ノータム〔名〕	notice to airmen (NOTAM)

能力 才能 適性〔名〕	ability, capability, power, competence
能力がある〔形〕	capable (a capable pilot 有能なパイロット)
除いては〔前〕	except (The weather is good except destination airport. 目的地空港を除いて、天気は良好である。)
望む 願う〔動〕	wish, desire, want
願望〔名〕	wish, desire
乗り込む〔動〕	board (on board radio equipment 機上の無線機器)
乗り継ぎ 通過〔名〕	transit (a transit passenger 通過客)

は

派遣〔名〕	dispatch (dispatcher ディスパッチャー 運航管理者)
派遣する〔動〕	dispatch
激しい〔形〕	violent (violent earthquake 激しい地震) turbulent (turbulent air 乱気流) strong (strong wind 強い風) intense (intense shock 激しいショック)
運ぶ〔動〕	carry (carry a bag バッグを運ぶ) transport (transport people 人々を運ぶ)
運搬 輸送〔名〕	transport transportation (air transportation 航空輸送)
始め 開始〔名〕	beginning (beginning of a day 一日の始め) start commencement (The pilot called out commencement of takeoff. パイロットは、離陸開始をコールアウトした。)
始める 開始する〔動〕	begin, start, initiate, commence (commence takeoff 離陸を開始する)

は

はっきりした 明瞭な〔形〕	clear (clear voice はっきりした声) distinct (distinct pronunciation 明瞭な発音)
はっきりと 明瞭に〔副〕	clearly, distinctly
発見〔名〕	discovery (important discovery 重要な発見)
発見する 検出する〔動〕	discover, find, detect (detect a stray frequency ずれた周波数を検出する)
発行 刊行物〔名〕	issue (the day of issue 発行日) publication (an official publication 公の刊行物)
発行する 出版する 発表する〔動〕	issue, publish
発振器〔名〕	oscillator (The oscillator generates a carrier wave. 発振器は、搬送波を発生させる。)
発振する 振動する〔動〕	oscillate
発信者〔名〕	originator (originator of the message 通報の発信者)
発生〔名〕	occurrence (occurrence of thunder 雷の発生) generation (generation of a shock wave 衝撃波の発生)
発生する 起る 生産する〔動〕	occur (Thunderstorms occur in winter as well. 雷雨は、冬にも発生する。) produce (生産する)
発生させる〔動〕	generate
発音〔名〕	pronunciation
発音する〔動〕	pronounce (pronounce clearly はっきり発音する)
発電機〔名〕	generator
話す〔動〕	speak (speak slowly ゆっくり話す) talk
話し方 スピーチ〔名〕	speech

は

幅広さ〔名〕	width (beam width 信号電波の幅 ビーム幅)
幅が広い〔形〕	wide (wide street 広い街路) broad (broad beam 幅の広い信号電波)
版〔名〕	version (original version 原版) edition (the first edition 初版)
範囲 程度〔名〕	range coverage (radar coverage レーダー電波の到達範囲) extent (to some extent ある程度までは)
半径〔名〕	radius (radius of a circle 円の半径)
反射〔名〕	reflection (an angle of reflection 反射角)
反射する 反省する〔動〕	reflect (reflected light 反射光) reflect (reflect on oneself 反省する)
搬送波〔名〕	carrier wave (A carrier wave is modulated by a signal wave. 搬送波は、信号波により変調される。)
反復 繰返し 復唱〔名〕	repetition (repetition by a pilot パイロットによる復唱)
反復する 繰り返す 復唱する〔動〕	repeat (The pilot repeated the clearance for verification. パイロットは、確認のためにクリアランスを復唱した。)
場所 地点 位置〔名〕	place (a cozy place 居心地の良い場所) spot area (parking area 駐機場) location
バンド 帯〔名〕	band (frequency band 周波数帯)
パイロット 操縦士〔名〕	pilot (qualified pilot 資格を持ったパイロット)
パイロット相互間の通信呼出用語	interpilot air-to-air communication
パターン〔名〕 場周経路〔名〕	pattern (traffic pattern 場周経路) (The controller vectored the aircraft

は

	to the traffic pattern. 管制官は、機を場周経路へ誘導した。)
パルス 波動〔名〕	pulse (pulse modulation (PM) パルス変調)

ひ

火 火災〔名〕	fire (fire extinguisher 消火装置)
控える 絶つ〔動〕	refrain (refrain from unnecessary transmission 不要な送信をしない)
比較〔名〕	comparison (comparison of endurance 耐久性の比較)
比較する〔動〕	compare (compare the cruise performance 巡航性能を比較する)
比較的に 割合に〔副〕	comparatively (comparatively easy 比較的容易)
低い〔形〕	low (low level 低高度)
飛行 航空 飛行家〔名〕	flight, aviation, aviator
飛行する〔動〕	fly
飛行計画〔名〕	flight plan
飛行計画記載の航路〔名〕	flight plan route
飛行高度〔名〕	flight level (FL), altitude
飛行場〔名〕	airport, aerodrome
飛行場監視レーダー〔名〕	airport surveillance radar (ASR)
飛行場情報放送業務〔名〕	automatic terminal information service (ATIS)
飛行情報区〔名〕	flight information region (FIR)
飛行の安全〔名〕	flight safety
飛行の進行〔名〕	flight progress
飛行の段階〔名〕	flight phase
飛行便名〔名〕	flight identification, flight number
非常に〔形〕	very (very high frequency (VHF) 超短波)
非常事態〔名〕	emergency (in an emergency 非常の時には)
非常用周波数〔名〕	emergency frequency

必須の 欠くことのできない〔形〕	essential (Weather information is essential for flight. 気象情報は、飛行にとって不可欠である。)
引っ張る〔動〕	pull (A pilot pulled the control column to liftoff. パイロットは、離陸するために操縦桿を引いた。)
必要な〔形〕 必要とする〔動〕	necessary (necessary equipment 必要な装備) need, require (required amount of fuel 必要な量の燃料)
否定的な〔形〕 否定する 否認する〔動〕	negative (negative response 否定的反応) deny (The pilot denied the dispatcher's proposal. パイロットは、ディスパッチャーの提案を否認した。)
人〔名〕	person (Fifty-four persons are on board. 54人の人が、搭乗している。)
等しい 同等の〔形〕	equal (equal rights 平等の権利) (equal time point (ETP) 同時間飛行可能点) equivalent (同等の)
非有償飛行〔名〕	non-revenue flight (訓練飛行 試験飛行など)
表現〔名〕 表現する〔動〕	expression (verbal expression 言葉による表現) express (Aircraft speed is expressed in IAS, TAS and GS. 航空機の速度は、対気速度、真対気速度及び対地速度で表現される。)
標識 信号〔名〕	beacon (beacon light 標識灯) (radio beacon 無線標識) signal (distress signal 遭難信号)
表示〔名〕 表示する〔動〕	indication, display indicate, display

表示器 計器〔名〕	display unit (表示器) indicator (計器)
デジタル表示〔名〕	digital display
標準 基準〔名〕	standard (safety standards 安全基準)
標準化〔名〕	standardization (worldwide standardization 世界規模の標準化)
標準の〔形〕	standard (standard terminology 標準用語)
標準化する〔動〕	standardize (standardize the operating procedures 操作手順を標準化する)
標準計器出発方式〔名〕	standard instrument departure (SID)
標準到着経路〔名〕	standard terminal arrival route (STAR)
標準方式〔名〕	standard procedure
表面〔名〕	surface (earth's surface 地球の表面)
地上の 水上の〔形〕	surface (surface movement 地上における移動)
品質〔名〕	quality (high quality 高品質)
頻繁な〔形〕	frequent (frequent occurrence 頻繁な発生)
頻繁に しばしば〔副〕	frequently (The crosswind component frequently exceeds the limit for landing. 横風成分が、着陸の限界をしばしば超える。)
尾部〔名〕	tail, tail section (tail hit 尾部の地面との接触)

ふ

ファイルする〔動〕	file (file the flight plan 飛行計画をファイルする)
風向〔名〕	wind direction (The wind shifted to the south. 風は、南に変わった。)
風速〔名〕	wind velocity (Change in wind velocity affects pilot's pitch control. 風速の変化は、パイロットのピッチコントロールに影響を及ぼす。)

賦課する 課する〔動〕	impose (MAYDAY imposes communication silence. MAYDAYは、通信の沈黙を課する。)
不確実な〔形〕	uncertain (uncertain estimate 不確実な推定)
付加的な 追加の〔形〕 添加物〔名〕	additional (Additional fuel is required. 追加燃料が必要である。) additive (anti-ice additives 防氷添加物)
不可能な〔形〕	impossible (impossible tasks 不可能な仕事)
付近 周辺 近所〔名〕	vicinity, neighborhood
復唱〔名〕 復唱する〔動〕	read back read back (The pilot read back the clearance. パイロットは、クリアランスを復唱した。)
復調〔名〕 復調する〔動〕	demodulation demodulate
含む〔動〕	contain (ATIS contains visibility and/or RVR. ATISは、視程及びまたはRVRを含んでいる。) include
符号 暗号〔名〕 符号化する〔動〕	code (numerical code 数字の符号) encode (encoder エンコーダー)
不平を言う がやがや騒ぐ〔動〕 〔squawk〕という語の慣習的用法	squawk (squawk the transponder トランスポンダーを作動させる) squawk (Pilots squawk about malfunctioning equipment by describing it in the flight log. パイロットは、装備の不具合について航空日誌に記述して報告する。)
不十分な〔形〕	insufficient (insufficient supply of fuel 燃料の供給不足) inadequate (The pilot's consideration to wind variation was inadequate.

	パイロットの風の変化に対する考慮は、不十分であった。） inefficient (Manual control is inefficient compared to automatic control. 手動制御は、自動制御に比べると能率が劣る。）
不適当な 不適切な〔形〕	inappropriate (inappropriate radar vectoring 不適切なレーダー誘導)
不法の 非合法的な〔形〕	unlawful (unlawful interference 不法な干渉)
不利な〔形〕 不利に〔副〕	adverse (adverse weather 悪天候) adversely (The wind adversely affected. 風は、不利に作用した。）
フレアー〔名〕 フレアーにする〔動〕	flare (start flare at 30 feet 30フィートでフレアーを開始する) flare (A pilot flares the aircraft to touchdown. パイロットは、接地するために機をフレアーさせる。）
部分〔名〕	part (main part 主要部分) portion (uncertain portion 不確実な部分) section (a section of the tower タワーの一部分)
分割（分割された）区分〔名〕 分割する〔動〕	division (There are three divisions in the tower. タワーには、三つの区分がある。） divide (The tower is divided into three sections. タワーは、三つの部分に分割されている。）
分離 離脱〔名〕 分離する 離脱する〔動〕	separation separate (Parallel runways are separated widely. 平行滑走路は、広く分離されている。）
分類〔名〕	classification

ふ

分類する〔動〕	classify (Radio waves are classified into eight frequency bands. 電波は、八つの周波数帯に分類されている。)
プッシュバック〔名〕	push back (push back clearance プッシュバックのクリアランス)
プッシュバックする〔動〕	push back (push back the aircraft 航空機をプッシュバックする)

へ

閉鎖 接近〔名〕	close, closure (closure rate 接近度)
閉鎖する〔動〕	close (The runway is closed for snow removal. 滑走路は、除雪のために閉鎖されている。)
変圧する〔動〕	transform (transformer 変圧器)
変調〔名〕	modulation (frequency modulation 周波数変調)
変調する〔動〕	modulate (modulator 変調器)
変化 変動〔名〕	change (change in weather 天気の変化) variation (wind variation 風の変動)
変化する 変動する〔動〕	change, vary (Wind direction varies. 風向が、変動する。)
別途〔副〕	separately (handle separately 別途取り扱う)
別の ほかの〔形〕	another (another task to do now 今やるべきもう一つの仕事)

ほ

方位 方位角（天文）〔名〕	azimuth (target azimuth by radar レーダーによるターゲットの方位)
方位 方位角（磁方位）〔名〕	bearing (bearing to the VOR station VOR局への磁方位)
方位の 指向性の〔形〕	directional (directional antenna 指向性アンテナ)
放棄する〔動〕	dump (fuel dump 燃料放棄) jettison (fuel jettison 燃料放棄)

報告〔名〕	report (a written report 書類による報告)
報告する〔動〕	report (A pilot must report completion of flight to a control unit. パイロットは、飛行の終了を管制機関に報告しなければならない。)
放射〔名〕	emission, radiation
放射する〔動〕	emit (emit radio waves 電波を放射する) radiate (The sun radiates light and heat. 太陽は、光と熱を放射する。)
方針 方策〔名〕	policy (the policy on flight safety 安全飛行の方策)
放送〔名〕	broadcast (ATIS is a broadcast of airport information. ATISは、飛行場情報の放送である。)
放送局〔名〕	broadcasting station
放送する〔動〕	broadcast (ATIS is continuously broadcasted. ATISは、連続的に放送されている。)
方法 手段〔名〕	method (a method of flight data analysis 飛行データ解析の方法) means (a means of communication 通信手段) manner (proper manner 適切な方法) way (the best way 最善の方法)
～のほかにも〔前〕	besides (strong wind besides rain 雨の上に強風)
他の〔形〕	other (some other time いつかまた)
保持する〔動〕	hold (a person holding a license 免許を保持している人)
保有する 保持する〔動〕	retain (The pressure accumulator retains regulated pressure well. 蓄圧器は、規正圧力をよく保持する。)

翻訳〔名〕	translation (This is a translation of ICAO regulation. これは、ICAO規則の翻訳である。)
翻訳する 解読する〔動〕	translate (SELCAL decoder translates the coded signal. SELCALのディコーダーは、コード化された信号を解読する。)
ボアスコープ〔名〕	bore-scope (bore-scope inspection ボアスコープを使った検査)
妨害 障害 じゃま 干渉〔名〕	disturbance, obstruction, interference
妨害する〔動〕	disturb, obstruct, interfere
防止する〔動〕	prevent (prevent from veer off the runway 滑走路からの逸脱を防ぐ)

ま

マイクロ波 極超短波〔名〕	microwave super high frequency (SHF)
撒き散らす〔動〕	scatter (The wind scattered the fog. 風が、霧を散らした。)
真下の〔前〕	directly below
間違い 誤り 失敗〔名〕	error, mistake, fault, failure
間違える〔動〕	make an error, make a mistake, fail
マッハ〔名〕	Mach (cruise at Mach .84 マッハ0.84で巡航する)
満足〔名〕	satisfaction
満足な〔形〕	satisfactory (satisfactory achievement 満足な成果)
満足に〔副〕	satisfactorily
満足させる〔動〕	satisfy (be satisfied with ～に満足である)

み

満たす〔動〕	meet (The weather conditions meet the minimum requirement for take-

	off. 気象状態は、離陸のための最低要件を満たしている。)
密集 渋滞〔名〕	congestion (traffic congestion 交通渋滞)
認める〔動〕	acknowledge (acknowledgement of receipt 受信証) recognize (The pilot recognized a hazardous target on the radar scope. パイロットは、レーダースコープ上に危険なターゲットを認めた。) admit (admit the error 誤りを認める)
認識〔名〕	recognition, acknowledgement
～とみなす ～と考える〔動〕	regard (DME is regarded as a set of interrogator and a transponder. DMEは、質問機と応答機のセットとみなされる。) consider
ミリ波〔名〕	extremely high frequency (EHF)

む

～に向かって〔前〕	toward (landing toward the setting sun 夕日に向かっての着陸)
無効な 効果的でない〔形〕	ineffective (ineffective consideration 無駄な考察)
無効にする〔動〕	annul (annul the previous clearance 先のクリアランスを無効にする)
無指向性無線標識〔名〕	non-directional radio beacon (NDB)
無視する〔動〕	disregard (disregard the proposal 提案を無視する)
矛盾 対立〔名〕	conflict (traffic conflict 相容れない交通の状態)
矛盾する 対立する〔動〕	conflict
無線航法〔名〕	radio navigation
無線航法施設〔名〕	radio navigation facility
無線周波数〔名〕	radio frequency
無線従事者〔名〕	radio operator

無線電話〔名〕	radio telephony
無線電話識別〔名〕	radio telephony designator
無線標識〔名〕	radio beacon
無線の〔形〕	radio (radio station 無線局) wireless (無線の)

め

明示する〔動〕	specify (specified terms 明示されている用語)
目の粗い〔形〕	coarse (coarse accuracy 目の粗い精度)
目盛を調整する〔動〕	calibrate (calibrate the altimeter 高度計の目盛を調整する)
免許 免許証〔名〕	certificate, license (radio operator's license 無線従事者の免許)
免除〔名〕 免除する〔動〕	exemption (exemption of flight plan 飛行計画の免除) exempt (Position report is exempted. 位置報告は、免除されている。)

も

モールス電信〔名〕	Morse telegraphy
盲目送信〔名〕	blind transmission (A pilot will send a blind transmission when the receiver has failed. パイロットは、受信機故障の場合には、盲目送信を行う。)
目視線〔名〕	line of sight
目的 意図 目標〔名〕 目的地 到着地〔名〕	purpose, object, objective, aim, intention destination (destination airport 目的地空港)
用いる 使う 適用する〔動〕	use (use of standard phraseology 標準用語の使用), employ (employ a new method 新しい方法を用いる), apply (apply new technique 新しい技術を適用する)

め
も

最も近い〔形〕	proximate (Stop at the stop line proximate to the runway. 滑走路に最も近い停止線で止まれ。)
～のもとにおいて〔前〕	under (under the control of the tower タワーの管制のもとにおいて)
もはや～でない〔形〕	no longer (The wind is no longer blowing. 風はもう吹いてはいない。)
～もまた〔副〕	as well (It's snowing and the wind is blowing as well. 雪が降っているし、風もまた吹いている。)

や

夜間効果〔名〕	night effect (Night effect of MF radio waves degrades communication quality. 中波の電波の夜間効果は、通信の質を低下させる。)
やり取り 交換〔名〕	exchange (an exchange of messages 通報の交換)
和らげる 弱める 減らす〔動〕	abate, reduce, decrease
騒音軽減方式〔名〕	noise abatement procedure

ゆ

有害な〔形〕	harmful (harmful interference 有害な混信)
有効性 効力〔名〕	effectiveness (effectiveness of a smoke detector 煙探知機の有効性)
有効な〔形〕	effective (HUD is an effective device. HUDは、有効な装置である。)
有視界気象状態〔名〕	visual meteorological condition (VMC)
有視界飛行方式〔名〕	visual flight rule (VFR)
有償飛行〔名〕	revenue flight
優先 優先権〔名〕	priority
優先順位〔名〕	priority, precedence
優先措置〔名〕	priority handling

誘導〔名〕	guidance (radio guidance 無線誘導) induction
誘導する〔動〕	guide, vector (radar vectoring レーダー誘導) induce (an induced current 誘導電流)
誘導路〔名〕	taxiway
有用な〔形〕	useful (useful information 有用な情報)
輸送 輸送機関〔名〕	transportation, transport
輸送する〔動〕	transport (transport by air 航空輸送)
航空輸送〔名〕	air transportation
許す 許可する 承諾する〔動〕	allow, permit, grant

よ

用意が整って〔形〕	ready (The pilot is ready to start takeoff. パイロットは、離陸開始の準備が出来ている。)
容易な〔形〕	easy (an easy task 容易な仕事)
容易に〔副〕	easily (The engine starts easily. エンジンは、容易にスタートできる。)
容易にする〔動〕	facilitate (A controller's advice facilitates pilot's flight management. 管制官の助言は、パイロットの飛行制御を容易にする。)
要件 必要条件〔名〕	requirement (The weather minima is a requirement for takeoff and landing. 最低気象条件は、離陸及び着陸の必要条件である。)
用語 述語 専門語〔名〕	term (technical term 専門語) phraseology, terminology
用語集〔名〕	glossary
洋上の 海洋の〔形〕	oceanic (oceanic climate 海洋性気候)
洋上管制〔名〕	oceanic control
洋上管制区域〔名〕	oceanic control area
洋上航空路監視レーダー〔名〕	oceanic route surveillance radar (ORSR)

要請 要求〔名〕 要請する〔動〕	request, demand request (request descent clearance 降下のクリアランスを要請する。)
要素 素子 要因〔名〕	element (antenna element アンテナの素子) factor (a factor of the problem 問題の要因)
要約 概要〔名〕 要約する〔動〕	summary (a summary of a report 報告の概要) summarize (summarize the report 報告を要約する)
予期 期待 予測〔名〕 予期する 期待する 予測する〔動〕	expectation, anticipation expect, anticipate (The pilot anticipated an icing condition. パイロットは、着氷状態を予測した。)
翼の前縁〔名〕	wing leading edge (The wing leading edge is protected from icing. 翼の前縁は、氷結に対して予防されている。)
横方向の〔形〕	lateral (The localizer provides lateral guidance signal. ローカライザーは、横方向の誘導シグナルを提供する。)
予知する〔動〕	foresee (The pilot foresaw control difficulties upon receipt of the weather information. パイロットは、気象情報を受取って操縦の困難を予知した。)
予定 スケジュール〔名〕	schedule
呼出し 呼出す〔動〕 呼出装置〔名〕	call call system (SELCAL is a call system added to the airborne receiver. SELCALは、機上の受信機に付加された呼出装置である。)

よ

呼出符号〔名〕	call sign (A call sign is assigned to each flight. 各フライトには、呼出符号が割当てられている。)
読取り可能な 聞取り可能な〔形〕	readable
読取り不可能な〔形〕	unreadable
読取り易さ〔名〕	readability (readability of the mesage 通報の読取り易さ)
ら	
乱気流〔名〕	turbulence (clear air turbulence 晴天乱気流)
り	
理解 会得 認識〔名〕	understanding, comprehension realization (認識)
理解する 会得する 認識する〔動〕	understand , comprehend realize (認識する)
率 割合〔名〕	rate (sink rate 沈下率)
略する〔動〕	abbreviate (abbreviate a word 語を略す)
理由〔名〕	reason (reason of delay 遅延の理由)
量 音量〔名〕	volume (turn up the receiver volume 受信機の音量を大きくする)
了解〔間頭詞〕	roger (了解を意味する通信用語)
旅行〔名〕	travel (travel by plane 航空旅行) trip, journey
旅行する〔動〕	travel light (軽装で旅行する) (make a trip 旅行する) (make a journey 旅行する)
利用可能な〔形〕	available (employ all available means すべての利用可能な手段を講じる)

離陸 離昇〔名〕	takeoff, lift off
離陸する 離昇する〔動〕	takeoff, lift off airborne (become airborne 離陸浮揚する)
離陸滑走〔名〕	takeoff roll
離陸滑走路〔名〕	takeoff runway
離陸許可〔名〕	takeoff clearance
離陸距離〔名〕	takeoff distance
離陸性能〔名〕	takeoff performance

る

類似の〔形〕	similar (similar call sign 類似のコールサイン)

れ

レーダー〔名〕	radar
レーダー装置〔名〕	radar equipment
レーダー覆域〔名〕	radar coverage
レーダー捕捉〔名〕	radar contact
レーダー誘導〔名〕	radar vectoring
レーダー誘導進入〔名〕	radar approach
例 例えば〔名〕	example (for example 例えば)
例外 除外〔名〕	exception (There's no exception. 例外は無い。)
連結する〔動〕	connect
連続 継続〔名〕	continuance (a continuance of good weather 好天の連続) series (a series of errors 誤りの連続)
連続する〔動〕	continue
連続的な〔形〕	continuous (a continuous rain ひっきりなしの雨)
連続的に〔副〕	continuously

ろ

ローカライザー〔名〕	localizer (LLZ) (Localizer is a part of ILS. ローカライザーは、ILSの一部分である。)
露点〔名〕	dew point

わ

割当 配置〔名〕	assignment, allotment, allocation (配置)
割当てる 配置する〔動〕	allot, allocate (配置する) assign (assigned SELCAL code 割当てられたSELCALコード)
割当て周波数〔名〕	assigned frequency

第3章

略語辞書

A

AAC	aeronautical administrative communication 運航業務通信　航空会社などの運航者が、INMARSATのVHFデータリンクを利用して航空機との間で行う運航業務に関する通信。
AC	alternating current 交流電流
ACARS	aircraft communication addressing and reporting system 空地データ通信システム　地上VHF通信局または衛星通信により、航空機と航空会社などの運航者との間の広範囲なデータ通信システム。
ACAS	airborne collision avoidance system 機上衝突防止装置 危険対象機の接近度に応じてTA、RAを発してパイロットに衝突防止を促すシステム。
ACC	area control center 航空路管制機関　航空路管制と進入管制を行う管制機関である。また、IFR機の管制承認を発行する。
ADF	automatic direction finder 自動方向探知機　NDB局の電波を受信し、局に対する相対方位を知る機上の受信装置。
ADS	automatic dependent surveillance 自動従属監視 レーダー監視空域外の洋上管制区で、自動的に航空機の位置監視を行うシステム。機上から送信される位置情報を管制スクリーン上に表示する。
AEIS	aeronautical en-route information service エンルート情報業務　飛行中の航空機に対して最新の気象情報を提供し、また、航空機から気象状態などの情報を受領して、他の航空機及び気象機関などに提供する業務。 AEISが取扱う情報の内NOTAMやPIREPなどは、VHFデータリンクでも提供される。
AFS	aeronautical fixed service 航空固定業務　地上局間で行う通信業務。管制機関、航空会社、情報処理施設などを含む通信網が行う業務。
AIC	aeronautical information circular 航空情報サーキュラー 内容や時間的な問題でAIPやNOTAMに適さない航空情報は、サーキュラーとして発行される。運航の安全、飛行の

方式などに関する規則などの説明を内容とする。

AIM aeronautical information manual 航空情報マニュアル
日本航空機操縦士協会発行の航空機運航に関する法規、基準、方式、技術などの解説書。

AIP aeronautical information publication 航空路誌
航空局発行の飛行場や保安施設に関する永続性のある情報を記載した刊行物。

AIREP air report 機上通報　国際航空路上の位置通報点及びその他必要に応じて行われるパイロットによる気象観測報告。

ALT altitude 高度　平均海面からの垂直距離または気圧高度。

AM amplitude modulation 振幅変調方式 AM方式
信号波の振幅に応じて搬送波の振幅を変調する方式。

AMS aeronautical mobile service　航空移動業務　航空局と航空機局、または、航空機相互間の無線通信業務。

AOC aeronautical operations control communication 運航管理通信　航空会社などの運航者が、INMARSATやVHFデーターリンクを利用して航空機との間で行う運航管理に関する通信。

APC aeronautical public communication 航空公衆通信
INMARSATを利用する機上の公衆通信。

ARP aerodrome reference point 飛行場標点　その飛行場を代表する地点。緯度、経度で表示し公表される。

ARSR air route surveillance radar 航空路監視レーダー
航空路上の交通管制を行うためのレーダー。

ASDE airport surface detection equipment 飛行場の滑走路、誘導路などにおける航空機および車両の交通管制を行うためのレーダー。

ASR airport surveillance radar 飛行場監視レーダー　飛行場周辺空域の出発および到着機の管制を行うためのレーダー。

ATC air traffic control 航空交通管制
航行中の航空機に対する交通管制。

ATIS	automatic terminal information service 飛行場情報放送業務　飛行場における諸般の状態、気象状態、保安施設の運用状態などの情報を定時的に連続提供する放送業務。
ATM	air traffic management 航空交通管理　従来の航空交通管制機能に加えて、衛星利用を含む航法、通信、監視システムによる航空交通管理の方式。
ATN	aeronautical telecommunication network 航空電気通信網　航空交通機関、運航者及び乗客に情報を提供する世界規模の通信網。
ATS	air traffic service 航空交通業務　航空交通のための管制業務、情報業務および緊急業務。福岡FIR内に提供される。

C

CAB	Civil Aviation Bureau 航空局
C/A code	coarse access code GPSの航法精度の粗い方。航法に適用するにはDGPSで精度を改善する。
CAL	civil aviation law 航空法
CAR	civil aviation regulation 航空法施行規則
CAT	clear air turbulence 晴天乱気流 雲の無い気流の中にある目視把握の出来ない乱気流で、ジェット気流の周辺に発生することが多い。
CAVOK	ceiling and visibility OK（雲高、視程ともに支障なし）の意の用語。
CCW	counterclockwise 反時計方向
CDI	course deviation indicator コース偏移指示計器 所望のコースと飛行コースの差異を指示する計器。
CNS	communication, navigation, surveillance 通信・航法・監視　航空交通管理システムの要素。
COP	changeover point 移管点　飛行の進行に伴って管制及び通信を次の施設に切替える点。通常各区間の中間点である。
CPDLC	controller pilot data-link communication 管制官とパイ

	ロットの間で行われる所定のフォーマットによるデーターリンク通信。洋上航空路におけるCPDLCによる位置通報により、ADSが可能となる。
CVR	cockpit voice recorder 操縦室内音声記録装置 操縦室内の音声を記録する装置。事故後回収されて、事故調査に使用される。
CW	clockwise 時計方向
Cb	cumulonimbus 積乱雲 発達した積乱雲は、運航に支障をきたす。

D

DA	decision altitude 決心高度 精密進入時の進入限界高度。平均海面からの高度。
DC	direct current 直流電流
DGPS	differential GPS ディファレンシャルGPS GPSのC/A codeの精度を向上させるシステム。MTSATやpseudo GPSが利用されている。
DH	decision height 決心高　精密進入時の進入限界高度のことであって、滑走路末端または接地帯の標高からの高度。
DME	distance measuring equipment 距離測定装置　機上のinterrogator（質問機）と地上のtransponder（応答機）の間における、電波による質問・応答により航空機と地上のDME装備との間の距離を測定する装置。DMEは、VOR局に併置される。測定距離は斜距離である。
DR	dead reckoning 推測航法 パイロットが、速度、針路、風向風速及び経過時間などを基に計算により推測位置を求める航法。

E

EHF	extremely high frequency ミリ波　30GHz~300GHzの電波。衛星通信やレーダーに使用される。
ELT	emergency locator transmitter 救命無線機 航空機に装備され、航空機が遭難した場合に、その送信の地点を探知させるための信号を自動的に送信するもの。

ETA	estimated time of arrival 到着予定時間
ETD	estimated time of departure 出発予定時間
ETP	equal time point 同時間飛行可能点　エンジン故障の場合に、目的地に飛行しても出発地に引き返しても同じ所要時間となる点。

F

FAF	final approach fix 最終進入フィックス 最終進入の開始点となるフィックス。
FDR	flight data recorder 飛行データ記録装置　飛行データを記録する装置。事故後回収されて、事故解析に使用される。
FIR	flight information region 飛行情報区 飛行情報業務および緊急業務を提供するためにICAOにより加盟国に割当てられた空域。わが国では福岡FIR。
FIS	flight information service 飛行情報業務　航空機の安全と円滑な運航に必要な情報を提供する業務。
FL	flight level フライトレベル　わが国では、14,000ft以上の飛行高度。
FM	frequency modulation　周波数変調 FM方式　信号波の周波数に応じて搬送波の周波数を変調する方式。
FREQ	frequency 周波数

G

G/A	go-around　着陸復行　着陸のための進入継続を決断した後、何らかの理由により行う復行。
GCA	ground controlled approach 地上誘導着陸方式 管制官が、ASRとPARを使用して、到着機を音声により着陸地点へと誘導する方式。
GHz	giga-Hertz ギガヘルツ
GLONASS	global orbiting navigation satellite system ロシアが運用する航法衛星。機能は米国のGPSと略同様。
GP	glide path グライドパス　電波または灯火により与えられる最終進入経路上の進入角。誘導電波は着陸接地帯ま

	で有効。
GPS	global positioning system 衛星航法システム　航法衛星からの電波を受信して位置を決定するシステム。
GPWS	ground proximity warning system 対地接近警報システム　機上に装備されている対地異常接近警報システム。
GS	glide slope グライドスロープ　最終進入経路上の進入角。
GS	ground speed 対地速度
	H
HDG	heading 機首方位
HF	high frequency 短波　3MHz~30MHzの周波数の電波　電離層反射で長距離伝播するので、長距離通信に適用される。
Hz	Hertz ヘルツ
	I
IAF	initial approach fix 初期進入開始点
IAS	indicated air speed 指示対気速度　航空機と大気との相対速度。ピトー静圧系からの圧力を速度計で指示する。系統による指示誤差は修正されていない。運航における速度の表示に適用される。
IATA	International Air Transportation Association 国際航空運送協会　主要航空会社が加盟する協会。運送面での協力を行う。
ICAO	International Civil Aviation Organization 国際民間航空機関　主要航空会社が加盟する国連の専門機関。航空機、飛行場施設、航路上の施設および運航方式など運航面での協力を行う。
IFR	instrument flight rule 計器飛行方式 航空機が、ATCクリアランスおよび管制官の指示に従って飛行する飛行方式。
ILS	instrument landing system 計器着陸方式　計器着陸装置　LLZおよびGPの電波の誘導を利用して行う精密進入着陸方式。LLZおよびGPの誘導電波を送信する滑走

路近くに設けられる装備。最終進入経路上の距離情報を提供するためのDMEやGPSも含まれる。

IMC instrument meteorological condition 計器飛行状態　視程や雲の状況から、飛行にとって視界不良の気象状態。

IMSO International Mobile Satellite Organization 国際移動通信衛星機構　この機構の下で世界の各社が、衛星通信を行っている。航空通信では、ATS用データ通信（CP-DLC利用のADS)、運航管理通信（AOC)、航空業務通信(AAC)および航空公衆通信（APC）が実施されている。

INS inertial navigation system 慣性航法装置　飛行に伴う加速度を検出し、航空機の姿勢、速度、距離を算出して航法データを提供する装置。高度情報は高度計による。

INTER-PILOT インターパイロット　飛行中他の航空機とのパイロット相互間の通信。通信の呼出しには、INTERPILOTを前置して行う。

IRS inertial reference system 慣性基準装置
飛行に伴う加速度をレーザージャイロによって検出し、計算過程を経て飛行、航法情報を提供する装置。

ISA international standard atmosphere 国際標準大気

ITU International Telecommunication Union 国際電気通信連合　電気通信に係わる国連の専門機関。

ITU-RR ITU Radio Regulations ITU無線通信規則　ITUで定めた無線通信に係わる規則。

J

JST Japan standard time 日本標準時
UTCより9時間進んでいる。

K

KHz kilo Hertz キロヘルツ

L

LAT latitude 緯度

LF low frequency 長波　30KHz~300KHzの電波。
NDBに適用。

LLWS	low level windshear 低高度ウィンドシヤー 最終進入コースまたは離陸及び初期上昇経路上にあるウィンドシヤー。
LLZ	localizer ローカライザー　ILSの平面誘導電波 滑走路中心線上でも有効。
LMM	locator middle marker　ILSミドルマーカーに併設されるNDB.
LOM	locator outer marker　ILSアウターマーカーに併設されるNDB.
LONG	longitude 経度
LRRA	low range radio altimeter 低高度電波高度計　定期旅客機用の電波高度計。0ft~2,500ftの絶対高度を指示する。精密進入着陸時の操縦操作の高度情報として使う。

M

MAY-DAY	MAY-DAY 航空機の遭難信号 遭難呼出しは、他の通信に対し絶対優先である。MAY-DAYは、他の無線局の沈黙を命じる用語である。
MDA	minimum descent altitude 最低降下高度 非精密進入を行う場合の進入限界高度。
MEA	minimum en-route altitude　最低経路高度　IFR機に対して設けられた航路上の最低高度。電波の到達距離及び地表または障害物からの距離を考慮して設定される。
MF	middle frequency 中波　300KHz~3MHzの電波。 NDBに適用。
MHA	minimum holding altitude 最低待機経路高度。 待機コースのための最低高度。
MHz	mega-Hertz メガヘルツ
MM	middle marker　ミドルマーカー　ILSグライドスロープ上の決心高(DA) に設置されるマーカービーコン。
MRA	minimum reception altitude　最低受信可能高度　関係する無線施設の信号を良好な状態で受信可能な最低高度。

MSL	mean sea level 平均海面
MTSAT	multi-functional transport satellite 運輸多目的衛星　CNS/ATMの中心核をなし、航空ミッションと気象ミッションを持つ多目的衛星。GPSの航法精度の改善、CPDLC利用のADSに適用されている。140°E、145°E、赤道上36,000kmにある二個の衛星。
MVA	minimum vectoring altitude 最低誘導高度 管制官が、レーダー誘導を行うに当たって航空機に指定できる最低高度。

N

NAV	navigation 航法
NAVAID	navigation aids 航法援助施設　飛行場施設、航路上の施設など。
NDB	non-directional radio beacon 無指向性無線標識　無指向性の電波を発する地上の施設。この電波を機上のADFで受信し地上局の方位を求める。
NOTAM	notice to airmen ノータム　運航関係者にとって不可欠な情報（保安施設の状態、運用方式など）をタイムリーに知らせるための通知。

O

OAC	oceanic area control center　洋上区域管制センター
OCA	oceanic control area　洋上管制区　福岡FIR内の洋上空域。
OM	outer marker アウターマーカー　ILSアプローチでグライドスロープをインターセプトする高度の真下に設置されるマーカービーコン。
ORSR	oceanic route surveillance radar 洋上航空路監視レーダー　洋上航空路の交通管制のためのレーダー。二次レーダーである。
OTR	oceanic transition route 洋上転移経路　陸上の無線施設と洋上管制区内のフィックスの間に設定される飛行経路。

P

P code	precision code GPSの精密精度。民間では使用しない。

PAN-PAN	PAN-PAN 航空機の緊急事態を表現する信号。緊急通信の順位は、遭難通信に次ぐ。他の通信は、緊急通信を妨げてはならない。緊急医療の場合は、PAN-PAN MEDICALの信号を使用する。
PAPI	precision approach path indicator 精密進入経路指示灯 進入するパイロットに適切な進入角を与える灯火。 接地帯付近に装備され、赤･白の灯火で指示する。
PAR	precision approach radar 精密進入レーダー GCAファイナルコントローラーが、最終進入コースにある航空機を誘導するために使用するレーダー。
PCA	positive control area 特別管制区　VFR機が、管制機関の許可を得ないで飛行することの出来ない公示された空域。
PIC	pilot in command 機長　飛行中、航空機の運航と安全に関して責任と権限を有するパイロット。
PIREP	pilot report 機上気象報告　国内航空路上で遭遇または観察した気象状況の報告。
PM	pulse modulation レーダーにおいて、基準パルス波を信号パルスで変調する変調方式。

Q

QFE	QFE 飛行場標高の大気圧による気圧高度計の規正値。 滑走路上での気圧高度計指示は0ftとなる。
QNE	QNE 平均海面上の較正大気圧による気圧高度計の規正値。29.92Hgまたは1013.2hPsを基準とする高度指示となる。
QNH	QNH 平均海面上の大気圧による気圧高度計の規正値。 飛行場または最寄の飛行経路の気圧による。該当する気圧を基準とする高度指示となる。該当する気圧は、管制機関またはATISにより提供される。

R

RA	radio altitude 電波高度計による高度。
RA	resolution advisory 回避情報 TCASが発生する衝突回避を促す指示。

RCC	rescue coordination center 救難調整本部 航空機の捜索救難情報を収集し、捜索救難の範囲を決定し、関連部署に捜索救難活動への出動を要請する機関。
RNAV	area navigation アールナブ　広域航法。 VOR、DMEあるいはGPSの方位と距離情報を利用して、飛行コースを定める航法の方式。
RVR	runway visual range 滑走路視距離　滑走路上の三点(接地点、中央、終点)で測定される視認距離。 離着陸気象要件の一つ。
RWY	runway 滑走路

S

SAR	search and rescue 捜索救難 遭難航空機に対する捜索救難活動。
SATCOM	satellite communication 衛星通信 衛星を利用した航空通信。
SELCAL	selective calling system セルコール　選択呼出装置 特定の航空機の無線電話呼出しに当たって、符号を用いて自動的に選択呼出しを行う装置。地上局のエンコーダーから発信する符号を機上のディコーダーが解読して呼出装置を作動させるシステム。
SHF	super high frequency マイクロ波　3GHz~30GHzの電波。衛星通信、レーダーに適用される。
SID	standard instrument departure 標準計器出発方式 IFR機が、離陸後滑走路から航空路またはトランジションに至るまでの所定の出発飛行方式。
SIGMET	significant meteorological information シグメット 悪天情報　運航に重大な影響をおよぼす悪天候情報。
SNOW-TAM	snow relating NOTAM 走行区域における降雪、雪、氷などに関するNOTAM。
SOP	standard operating procedure 標準操作手順　無線機器の操作については、ICAOにより標準化された手順が、規定されている。

SRA	surveillance radar approach 捜索レーダーアプローチ ASRを使用してコースと高度を誘導する非精密進入方式。
SSR	secondary surveillance radar 二次監視レーダー ASRやARSRに併設されていて、特定の航空機を、一次レーダースクリーン上で識別することを可能にする二次レーダーである。
STAR	standard terminal arrival route 標準到着経路 航空路上のフィックスから最終進入フィックスまでの間を、IFR機が飛行するために設定された飛行経路。

T

TA	traffic advisory 対抗機位置情報 TCASが発生する近接機の位置情報。
TACAN	tactical air navigation aid　タカン 航空機に距離と方位の情報を提供する航法援助施設。
TAS	true air speed 真対気速度　撹乱されない大気に対する航空機の速度。飛行計画、航法に適用される。
TCA	terminal control area ターミナル管制区　進入管制区の中でVFR機が多く飛行する空域を指定して、VFR機に対してもレーダーサービスを提供する空域。
TCAS	traffic alert and collision avoidance system 航空機衝突防止装置　ACASともいう。機上装備のATCトランスポンダーの応答信号から所定範囲内にいる他の航空機の接近度を検出し、接近度によりTAおよびRAの衝突接近情報または回避指示をパイロットに与える。
TRU	transformer rectifier unit 航空機の電力供給システム内にあって、AC電圧を調節し、ACをDCに変換する装置。
TWY	taxiway 誘導路　駐機場から滑走路までの間の航空機が移動するための通路。

U

UHF	ultra high frequency 極超短波　300MHz~3GHzの電波。 ILSのGP、DME、管制用レーダーに適用されている。
UTC	coordinated universal time　協定世界時　世界標準時(GMT)

V

VAA	volcanic ash advisory 火山灰注意報
VASIS	visual approach slope indicator system 進入角指示灯
VFR	visual flight rule 有視界飛行方式 有視界飛行状態を維持して飛行する飛行方式。
VHF	very high frequency 超短波　30MHz~300MHzの電波。航空管制通信に適用されている。
VLF	very low frequency 超長波　3KHz~30KHzの電波。
VMC	visual meteorological condition 有視界気象状態　視程および雲から航空機までの距離が、所定値以上の気象状態。
VOLMET	voice meteorological broadcast ボルメット　HFを使用して音声で広域に対して飛行場の気象状況を放送する業務。
VOR	VHF omni-directional radio range 超短波全方向式無線標識　基準位相と可変位相のVHF電波を発信する地上の標識局。機上の受信機が位相差を検出し、航空機の局に対する磁方位を求める。
VORTAC	VOR-TACAN　VORとTACANを併設した無線標識。 方位と距離の情報を提供する。

W

WILCO	will comply (通報を了解しました。) (それに従います。) を意味する通報用の略語。

航空無線通信士　英語簡易辞書　（電略コカ）

平成22年２月１日　第１版第１刷発行
令和３年３月12日　第２版第１刷発行

発 行 所　一般財団法人情報通信振興会
郵便番号　170-8480
東京都豊島区駒込２-３-10
電　話　（03）3940-3951（販売）
（03）3940-8900（編集）
ＦＡＸ　（03）3940-4055
ＵＲＬ　https://www. dsk. or. jp/

印刷　株式会社エム．ティ．ディ

ISBN978-4-8076-0937-6　C3555　¥1800E